My Book

AF887786

This book belongs to

Name: ______________________________

Cover Design by :
Gowri Vemuri

First Edition :
April, 2020

Author :
Gowri Vemuri

Editor :
Ritvik Pothapragada

Questions: mathknots.help@gmail.com

Dedication

This book is dedicated to:
My Mom, who is my best critic, guide and supporter.
To what I am today, and what I am going to become tomorrow,
is all because of your blessings, unconditional affection and support.

This book is dedicated to the
strongest women of my life ,
my dearest mom
and
to all those moms in this universe.

G.V.

<u>Specialized High School Admission Test (SHSAT</u>) scores determines the admission to any/all of New York City's eight Specialized High Schools as given below:

(1)The Bronx High School of Science
(2) The Brooklyn Latin School
(3) Brooklyn Technical High School
(4) High School for Mathematics, Science and Engineering at City College
(5) High School of American Studies at Lehman College
(6) Queens High School for the Sciences at York College
(7) Staten Island Technical High School
(8) Stuyvesant High School

This book only covers Math section of the test. Students can choose Math section first or ELA
Total time for both sections is 180 minutes
It is up to the student to decide which section you take first!

(1) Work carefully but keep moving at a comfortable pace.
(2) Do not spend more than a couple minutes on any one question.
(3) Remember to note for the questions you want to review time permitting towards the end
(4) Bubble your answers on the answer sheet as you work through each question.
(5) Make sure answers are completely filled in.
(6) All answers must be marked on the answer sheet provided during the test time.

Remember: Students will not receive any extra time to bubble the answers after 180 minutes of time is over

(7) All correct answers will receive score.
(8) There is no negative marking
(9) Do not leave any questions unanswered. Do a wise guess by eliminating the given answer choices that are definitely wrong. Usually, we can eliminate 2 choices with careful observation.
(10) If you finish the test, then recheck the questions that you wanted to review. also make sure all questions are answered.
(11) total duration of the test is 180 minutes.

<u>Math Test details:</u>

The Math section consists of word problems and computational questions in either a grid-in or multiple-choice format. There are five grid-in questions and 52 multiple-choice questions. The Math questions involve application of mathematical skills, mathematical terms, and general concepts that align to New York State Learning Standards for Mathematics.
The questions are based on the application of the core concepts to complex, multi-step problems.
If it is taking too long or you are unsure of how to answer a question, skip it and return to it after answering the remaining questions.

<u>SCORE CALCULATION</u>

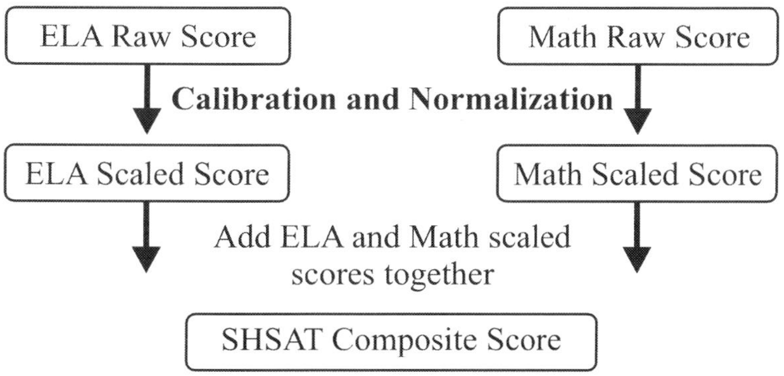

Fractions Notes

Fractions are part of a whole. It is also an expression representing quotient of two quantities.

Example 1 : $\frac{2}{4}$, $\frac{1}{4}$

Fraction can also be represented as ratios.

Example 2 : 1 : 2 or $\frac{1}{2}$

3 : 7 of $\frac{3}{7}$

Adding simple fractions, follow the below steps :

#1 : Fractions with like denominators can be added by adding their numerators.

Example 3 : $\frac{3}{5} + \frac{1}{5} = \frac{3+1}{5} = \frac{4}{5}$

#3 : To add fractions with unlike denominators, convert the fractions to equivalent fractions with like denominators and follow #1. To convert them into equivalent fractions you can multiply the numerator and denominator with a common factor for each of the fractions to be added separately and then add the fractions.

Method 1 :

Example 4 : $\frac{2}{5} + \frac{1}{2}$

Step 1 : $\frac{2}{5} = \frac{2 \text{ X } 2}{5 \text{ X } 2} = \frac{4}{10}$

Step 2 : $\frac{1}{2} = \frac{1 \text{ X } 5}{2 \text{ X } 5} = \frac{5}{10}$

Step 3 : $\frac{2}{5} + \frac{1}{2} = \frac{4}{10} + \frac{5}{10} = \frac{4+5}{10} = \frac{9}{10}$

$\frac{2}{5} + \frac{1}{2} = \frac{9}{10}$

Another method to add fractions with unlike denominators is by finding the **Least Common Multiple** (LCM) of the denominators and then follow the steps as described below in same order.

Method 2 :

Example 5 : $\frac{5}{12} + \frac{1}{8}$

Step 1 : Find the LCM of the denominators 12 , 8

$$\begin{array}{r|l} 2 & 12\,,\,8 \\ \hline 2 & 6\,,\,4 \\ \hline 3 & 3\,,\,2 \\ \hline 2 & 1\,,\,2 \\ \hline & 1\,,\,1 \end{array}$$

LCM = 2 X 2 X 3 X 2 = 24

Step 2 : Converting $\frac{5}{12}$ into an equivalent fraction with the denominator as 24 (LCM).

Divide the LCM value with the number in the denominator of the fraction to obtain the common factor.

$$\begin{array}{r} 2 \\ 12\overline{)\,24} \\ -24 \\ \hline 0 \end{array}$$

Common factor obtained is 2

$$\frac{5}{12} = \frac{5 \text{ X } 2}{12 \text{ X} 2} = \frac{10}{24} \; ; \; \frac{5}{12} = \frac{10}{24}$$

Step 3 : Converting $\frac{1}{8}$ into an equivalent fraction with the denominator as 24 (LCM).

Divide the LCM value with the number in the denominator of the fraction to obtain the common factor.

$$\begin{array}{r} 3 \\ 8\overline{)\,24} \\ -24 \\ \hline 0 \end{array}$$

Common factor obtained is 3

$$\frac{1}{8} = \frac{1 \text{ X } 3}{8 \text{ X } 3} = \frac{3}{24} \; ; \; \frac{1}{8} = \frac{3}{24}$$

Step 4 : Substitute the equivalent fractions obtained in step 2 and 3.

$$\frac{5}{12} + \frac{1}{8} = \frac{10}{24} + \frac{3}{24}$$

Step 5 : The denominators of both fractions are same (Like denominators). Add the numerators.

$$\frac{10 + 3}{24} = \frac{13}{24}$$

> **Note :** To add more than two fractions, repeat step 2 or step 3 for each of the fractions to convert them into equivalent fractions and then proceed with step 4 and 5.

$$\frac{5}{12} + \frac{1}{8} = \frac{13}{24}$$

Subtracting simple fractions, follow the below steps :

#1 : Fractions with like denominators can be subtracted by subtracting their numerators.

Example 6 : $\frac{3}{5} - \frac{1}{5} = \frac{3 - 1}{5} = \frac{2}{5}$

#2 : To subtract fractions with unlike denominators, convert the fractions to equivalent fractions with like denominators and follow #1. To convert them into equivalent fractions you can multiply the numerator and denominator with a common factor for each of the fractions to be subtracted separately and then subtract the fractions.

Method 1 :

Example 7 : $\frac{3}{5} - \frac{1}{2}$

Step 1 : $\frac{3}{5} = \frac{3 \text{ X } 2}{5 \text{ X } 2} = \frac{6}{10}$

Step 2 : $\frac{1}{2} = \frac{1 \text{ X } 5}{2 \text{ X } 5} = \frac{5}{10}$

Step 3 : $\frac{3}{5} - \frac{1}{2} = \frac{6}{10} - \frac{5}{10} = \frac{6 - 5}{10} = \frac{1}{10}$

$$\frac{3}{5} - \frac{1}{2} = \frac{1}{10}$$

Another method to subtract fractions with unlike denominators is by finding the **Least Common Multiple** (LCM) of the denominators and then follow the steps as described below in same order.

Method 2 :

Example 8 : $\frac{5}{12} - \frac{1}{8}$

Step 1 : Find the LCM of the denominators 12 , 8

$$\begin{array}{r|l} 2 & 12 , 8 \\ \hline 2 & 6 , 4 \\ \hline 3 & 3 , 2 \\ \hline 2 & 1 , 2 \\ \hline & 1 , 1 \end{array}$$

LCM = 2 X 2 X 3 X 2 = 24

Step 2 : Converting $\frac{5}{12}$ into an equivalent fraction with the denominator as 24 (LCM).

Divide the LCM value with the number in the denominator of the fraction to obtain the common factor.

$$\begin{array}{r} 2 \\ 12 \overline{)\ 24} \\ -\ 24 \\ \hline 0 \end{array}$$

Common factor obtained is 2

$$\frac{5}{12} = \frac{5 \text{ X } 2}{12 \text{ X} 2} = \frac{10}{24} \; ; \; \frac{5}{12} = \frac{10}{24}$$

Step 3 : Converting $\frac{1}{8}$ into an equivalent fraction with the denominator as 24 (LCM).

Divide the LCM value with the number in the denominator of the fraction to obtain the common factor.

$$\begin{array}{r} 3 \\ 8 \overline{)\ 24} \\ -\ 24 \\ \hline 0 \end{array}$$

Common factor obtained is 3

$$\frac{1}{8} = \frac{1 \text{ X } 3}{8 \text{ X } 3} = \frac{3}{24} \; ; \; \frac{1}{8} = \frac{3}{24}$$

Step 4 : Substitute the equivalent fractions obtained in step 2 and 3.

$$\frac{5}{12} - \frac{1}{8} = \frac{10}{24} - \frac{3}{24}$$

Step 5 : The denominators of both fractions are same (Like denominators). Subtract the numerators.

$$\frac{10 - 3}{24} = \frac{7}{24}$$

Note : To Subtract more than two fractions, repeat step 2 or step 3 for each of the fractions to convert them into equivalent fractions and then proceed with step 4 and 5.

$$\frac{5}{12} - \frac{1}{8} = \frac{7}{24}$$

Adding Mixed numbers :

#1 : Add the whole numbers together.

#2 : Add the fractional parts together. (Find a common denominator if necessary)
(Follow the steps as described in the previous pages)

#3 : Write the whole number obtained from step 1 and the fraction obtained from step 2.

#4 : If the fractional part is an improper fraction, change it to a mixed number.
Add the whole part of the mixed number to the original whole numbers.
Rewrite the fraction in the lowest possible value.

Example 9 : $3\frac{2}{7} + 5\frac{1}{7}$

Step 1 : Add the whole part 3 from $3\frac{2}{7}$ to the whole part 5 from $5\frac{1}{7}$

$3 + 5 = 8$

Step 2 : Add the fractional part $\frac{2}{7}$ from $3\frac{2}{7}$ to the fractional part $\frac{1}{7}$ from $5\frac{1}{7}$

$$\frac{2}{7} + \frac{1}{7} = \frac{2 + 1}{7} = \frac{3}{7}$$

Step 3 : $3\frac{2}{7} + 5\frac{1}{7} = 8\frac{3}{7}$

Example 10 : $4\frac{1}{6} + 5\frac{5}{6}$

Step 1 : Add the whole part 4 from $4\frac{1}{6}$ to the whole part 5 from $5\frac{5}{6}$

$4 + 5 = 9$

Step 2 : Add the fractional part $\frac{1}{6}$ from $4\frac{1}{6}$ to the fractional part $\frac{5}{6}$ from $5\frac{5}{6}$

$\frac{1}{6} + \frac{5}{6} = \frac{1+5}{6} = \frac{6}{6}$ (Numerators are added for fractions from like denominators)

Step 3 : $4\frac{1}{6} + 5\frac{5}{6} = 9\frac{6}{6} = 10$

($\frac{6}{6}$ = one whole part, add this one whole part to 9 making it equal to 10)

Example 11 : $7\frac{1}{2} + 5\frac{3}{20}$

Step 1 : Add the whole part 7 from $7\frac{1}{2}$ to the whole part 5 from $5\frac{3}{20}$

$7 + 5 = 12$

Step 2 : Add the fractional part $\frac{1}{2}$ from $7\frac{1}{2}$ to the fractional part $\frac{3}{20}$ from $5\frac{3}{20}$

$\frac{1}{2} + \frac{3}{20}$ (The fractions has unlike denominators)

Finding the LCM of 2 and 20

2	2 ,20
2	1 ,10
5	1 , 5
	1 , 1

LCM of 2 and 20 = 2 X 2 X 5 = 20

Step 3 : Let's make $\frac{1}{2}$ as an equivalent fraction with a denominator of 20.

$$\begin{array}{r} 10 \\ 2\overline{)\,20} \\ -20 \\ \hline 0 \end{array}$$

Common factor is 10

$$\frac{1}{2} = \frac{1 \text{ X } 10}{2 \text{ X } 10} = \frac{10}{20}$$

Step 4 : The fractional part $\frac{3}{20}$ of $5\frac{3}{20}$ has the same common denominator as 20.

We do not need to convert $\frac{3}{20}$ into another equivalent fraction.

Remember : The fractions can vary from problem to problem and students need to follow step 3 for all the fractional parts to convert them to equivalent fractions.

Step 5 : $\frac{1}{2} + \frac{3}{20} = \frac{10}{20} + \frac{3}{20} = \frac{10+3}{20} = \frac{13}{20}$

Step 6 : $7\frac{1}{2} + 5\frac{3}{20} = 8\frac{13}{20}$

Subtracting Mixed numbers :

#1 : Subtract the whole numbers together.

#2 : Subtract the fractional parts together. (Find a common denominator if necessary) (Follow the steps as described in the previous pages)

#3 : Write the whole number obtained from step 1 and the fraction obtained from step 2.

#4 : If the fractional part is an improper fraction, change it to a mixed number. Add the whole part of the mixed number to the whole number obtained in step 1. Rewrite the fraction in the lowest possible value.

Example 12 : $9\frac{2}{7} - 5\frac{1}{7}$

Step 1 : Subtract the whole part 5 from $5\frac{1}{7}$ from the whole part 9 from $3\frac{2}{7}$

$9 - 5 = 4$

Step 2 : Subtract the fractional part $\frac{1}{7}$ from $5\frac{1}{7}$ from the fractional part $\frac{2}{7}$ from $3\frac{2}{7}$

$\frac{2}{7} - \frac{1}{7} = \frac{2-1}{7} = \frac{1}{7}$

Step 3 : $9\frac{2}{7} - 5\frac{1}{7} = 4\frac{1}{7}$

Example 13 : $7\frac{5}{6} - 5\frac{1}{6}$

Step 1 : Subtract the whole part 5 from $5\frac{1}{6}$ from the whole part 7 from $7\frac{1}{6}$

7 - 5 = 2

Step 2 : Subtract the fractional part $\frac{1}{6}$ from $5\frac{1}{6}$ to the fractional part $\frac{5}{6}$ from $7\frac{5}{6}$

$\frac{5}{6} - \frac{1}{6} = \frac{5-1}{6} = \frac{4}{6}$ (Numerators are added for fractions from like denominators)

Step 3 : $7\frac{5}{6} - 5\frac{1}{6} = 2\frac{4}{6} = 2\frac{2}{3}$

($\frac{4}{6} = \frac{2}{3}$ Equivalent fractions)

Example 14 : $7\frac{1}{2} - 5\frac{3}{20}$

Step 1 : Subtract the whole part 5 from $5\frac{3}{20}$ from the whole part 7 from $7\frac{1}{2}$

7 - 5 = 2

Step 2 : Subtract the fractional part $\frac{3}{20}$ from $5\frac{3}{20}$ to the fractional part $\frac{1}{2}$ from $7\frac{1}{2}$

$\frac{3}{20} - \frac{1}{2}$ (The fractions has unlike denominators)

Finding the LCM of 2 and 20

$$\begin{array}{r|l} 2 & 2,20 \\ \hline 2 & 1,10 \\ \hline 5 & 1,5 \\ \hline & 1,1 \end{array}$$

LCM of 2 and 20 = 2 X 2 X 5 = 20

Step 3 : Let's make $\frac{1}{2}$ as an equivalent fraction with a denominator of 20.

$$\begin{array}{r} 10 \\ 2\overline{)\,20} \\ -20 \\ \hline 0 \end{array}$$

Common factor is 10

$\frac{1}{2} = \frac{1 \times 10}{2 \times 10} = \frac{10}{20}$

Step 4 : The fractional part $\frac{3}{20}$ of $5\frac{3}{20}$ has the same common denominator as 20.

We do not need to convert $\frac{3}{20}$ into another equivalent fraction.

Remember : The fractions can vary from problem to problem and students need to follow step 3 for all the fractional parts to convert them to equivalent fractions.

Step 5 : $\frac{1}{2} - \frac{3}{20} = \frac{10}{20} - \frac{3}{20} = \frac{10 - 3}{20} = \frac{7}{20}$

Step 6 : $7\frac{1}{2} - 5\frac{3}{20} = 2\frac{7}{20}$

Example 15 : $5\frac{1}{7} - 3\frac{3}{7}$

Step 1 : Subtract the whole part 3 from $3\frac{3}{7}$ from the whole part 5 from $5\frac{1}{7}$

5 - 3 = 2

Step 2 : Subtract the fractional part $\frac{3}{7}$ from $3\frac{3}{7}$ from the fractional part $\frac{1}{7}$ from $5\frac{1}{7}$

We cannot subtract $\frac{3}{7}$ from $\frac{1}{7}$

So we need to rewrite the fraction $5\frac{1}{7}$

$5\frac{1}{7} = 4\frac{8}{7}$ (Remember in this fraction one whole part equals to seven. so when we take one whole part into the fraction form we need to add 7 to the value in the numerator which equals to 7 + 1 = 8)

Step 3 : Repeat step 1

$5\frac{1}{7} - 3\frac{3}{7} = 4\frac{8}{7} - 3\frac{3}{7}$

Subtract the whole part 3 from $3\frac{3}{7}$ from the whole part 4 from $4\frac{8}{7}$

4 - 3 = 1

Step 4 : Subtract the fractional part $\frac{3}{7}$ from $3\frac{3}{7}$ from the fractional part $\frac{8}{7}$ from $4\frac{8}{7}$

$\frac{8}{7} - \frac{3}{7} = \frac{8-3}{7} = \frac{5}{7}$

Step 5 : $5\frac{1}{7} - 3\frac{3}{7} = \frac{5}{7}$

Multiplying Fractions :

#1 : Verify if the fractions are in lowest possible values. If not convert them into lowest possible values.

#2 : Using cross simplification method simplify the fractions, meaning a numerator can be simplifies with a denominator only and vice versa.

#3 : Do not cross simplify numerator with a numerator value and denominator with a denominator value

#4 : Multiply the numerator with the remaining numerator values and the denominator with the denominator values

Remember : "Top times the top over the bottom times the bottom".
All the answers must be written in simplest form.

Example 16 : $\frac{6}{15} \times \frac{3}{10}$

$\frac{\cancel{6}^{2}}{\cancel{15}_{5}} \times \frac{3}{10} = \frac{\cancel{2}^{1}}{5} \times \frac{3}{\cancel{10}_{5}} = \frac{1}{5} \times \frac{3}{5}$

$= \frac{1 \times 3}{5 \times 5} = \frac{3}{25}$

Multiplying Mixed Numbers :

#1 : To multiply mixed numbers, convert them to improper fractions.

Converting mixed number to improper fractions :
Multiply the denominator of the fraction to the whole part and then add the product to the numerator.

Example 17 : $2\frac{1}{3} = \frac{2 \times 3 + 1}{3} = \frac{6 + 1}{3} = \frac{7}{3}$

#2 : Verify if the fractions are in lowest possible values. If not convert them into lowest possible values.

#3 : Using cross simplification method simplify the fractions, meaning a numerator can be simplifies with a denominator only and vice versa.

#4 : Do not cross simplify numerator with a numerator value and denominator with a denominator value

#5 : Multiply the numerator with the remaining numerator values and the denominator with the denominator values

Remember : "Top times the top over the bottom times the bottom".
All the answers must be written in simplest form.
All improper fractions must be change back to mixed numbers.

Note : MULTIPLICATION CAN BE WRITTEN WITH THE SYMBOLS X OR **.** IN BETWEEN, .

Dividing Fractions :

#1 : To divide fractions, convert the division problem into a multiplication problem. Do this by multiplying the first fraction by the reciprocal of the second fraction. In other words convert the division to multiplication and interchange the numerator and denominator of the second fraction.

Remember : "When two fractions we divide, flip the second and multiply."

Note : Don't forget to check for "cross simplification" when multiplying.
All answers must be written in simplest form.
All improper fractions must be change back to mixed numbers.

Example 18 : $\frac{6}{15} \div \frac{3}{10}$

$$\frac{6}{15} \div \frac{3}{10} = \frac{\overset{2}{\cancel{6}}}{\underset{5}{\cancel{15}}} \times \frac{10}{3} = \frac{2}{\underset{1}{\cancel{5}}} \times \frac{\overset{2}{\cancel{10}}}{3} = \frac{2}{1} \times \frac{2}{3}$$

$$= \frac{2 \times 2}{1 \times 3} = \frac{4}{3}$$

Dividing Mixed Numbers :

#1 : To divide mixed fractions, first change them to improper fractions.

#2 : To divide fractions, convert the division problem into a multiplication problem. Do this by multiplying the first fraction by the reciprocal of the second fraction. In other words convert the division to multiplication and interchange the numerator and denominator of the second fraction.

Remember : "When two fractions we divide, flip the second and multiply."

Note : Don't forget to check for "cross simplification" when multiplying.
All answers must be written in simplest form.
All improper fractions must be change back to mixed numbers.

Example 19 : $2\frac{6}{15} \div 5\frac{3}{10}$

$$2\frac{6}{15} \div 5\frac{3}{10} = \frac{2 \times 15 + 6}{15} \div \frac{5 \times 10 + 3}{10} = \frac{30 + 6}{15} \div \frac{50 + 3}{10}$$

$$= \frac{30+6}{15} \div \frac{50+3}{10}$$

$$= \frac{36}{15} \div \frac{53}{10}$$

$$= \frac{36}{15} \times \frac{10}{53}$$

(Remember when we change division to multiplication, flip the second fraction)

$$= \frac{\overset{12}{\cancel{36}}}{\underset{5}{\cancel{15}}} \times \frac{10}{53}$$

$$= \frac{12}{\underset{1}{\cancel{5}}} \times \frac{\overset{2}{\cancel{10}}}{53}$$

$$= \frac{12}{1} \times \frac{2}{53}$$

$$= \frac{12 \times 2}{1 \times 53}$$

$$= \frac{24}{53}$$

$$2\frac{6}{15} \div 5\frac{3}{10} = \frac{24}{53}$$

Decimals notes

The word standard means regular. Numbers in standard form are whole numbers or natural numbers.

Example : The number "six hundred twenty five" in standard form is 625.

To name a decimal from its standard form, follow these steps :

1. Name the number in front of the decimal. (Do not include the word "and").
2. The word "and" is used for the decimal point.
3. The number in the decimal part is similar to the number in front of the decimal.
4. Name the last place value given (of the digit farther to the right).

Rounding Decimals to a Given Place Value:

To round a decimal number to a given place value, look at the digit to the right of the desired place value and follow the rounding rules:

"5 and above, give it a shove! 4 or below, leave it alone!"

(The number in the desired place value gets "bumped up" to the next consecutive value if the digit to the right of it is 5 or more. The number in the desired place value does not change if the digit to the right of it has a value of 5 or less.)

The purpose of rounding is to provide *an* estimate and get an approximate value. Rounding involves losing some accuracy.

Example : If 5,953 people attend a Soccer game. We can approximately say 6,000 people watched the game.

Comparing and Ordering Decimals:

To order the given decimals, compare the digits of all decimals according to their place value.

Tip : Line the numbers up vertically according to their place value and compare from the left to the right. When comparing like place values from the left, the number with the higher digit is the larger number.

Also, added zeros to the right of a decimal number does not change the value of the decimal.

For example, 0.8 = 0.80 = 0.800 = 0.8000 = 0.80000

Naming/Reading Decimal :

Read the number to the left of the decimal. Say the word “and” for the decimal place. Read the number to the right of the decimal as you would read it if the number were on the left. End the name with the place value of the digit that is furthest to the right.

Converting Between Fractions and Decimals :

Fractions and decimals represent a certain “part of a whole”, all fractions can be written as decimals and terminating/repeating decimals can be written as fractions.

Multiplying Decimals :

To multiply decimals, first ignore the decimals and simply multiply the digits. Then count the total number of spaces from the right to the decimal (or digits to the right of the decimal) of both numbers; place the decimal that number of spaces from the right into your answer.

Dividing Decimals:

Divide the decimals as a whole numbers ignoring the decimals.

1. Count the number of digits of the dividend after the decimal.
2. Count the number of digits of the divisor.
3. Subtract the number obtained in step 1 from the number obtained instep 2.
4. Count the number of digits from the right to the number obtained in step 3.
5. Place the decimal point.

Integers notes

Integers are a group , or set of numbers that consist of "whole numbers and their opposites"

1. Natural numbers and whole numbers are subset of integers.
2. The set does not include fractions or decimals.
3. The set includes positive and negative numbers.
4. Integers include : $-\infty$..... , -5 , -4 , -3 , -2 , -1 , 0 , 1 , 2 , 3 , 4 , 5 ...... $+\infty$
5. Integers greater than zero are called positive integers.
6. Integers less than zero are called negative integers.
7. Zero is neither negative nor positive.
8. Negative integers are the numbers to the left of 0.
   Example : -5 , -4 , -3
9. Negative numbers have a negative (-) sigh in front of the number.
10. Positive integers are the numbers to the right of 0.
11. Positive numbers do not require the + sign in front.
12. If a number has no sign , it is a positive number.
    Example : 2 , 3 , 4 , 10 , 20
13. Negative numbers are frequently used in measurements.
    Example : To measure temperatures , depth etc
    4^0 C below zero degree celsius is represented as -4^0
    100 ft below sea level is represented as -100 ft.
14. Arrows on a number line represent the numbers continuing for ever.
15. Positive numbers are represented on the right side of zero on the number line.
16. Negative numbers are represented on the left side of zero on the number line.
17. Number are placed at equal intervals on the number line. Not necessarily one unit.

Integers can be represented on a number line as below

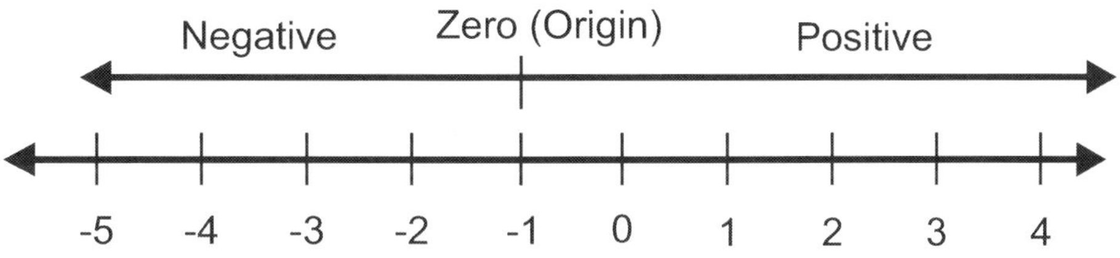

Absolute value :

The number line can be used to find the absolute value. The absolute value of an integer is the distance the number is from zero on the number line.

The absolute value of 2 is 2. Using the number line , 2 is a distance of 2 to the right of zero.The absolute value of -2 is also 2. Again using the number line , the distance from -2 to zero is 2. A measure of distance is always positive.

The symbol for absolute value of any number , x , is | x |.

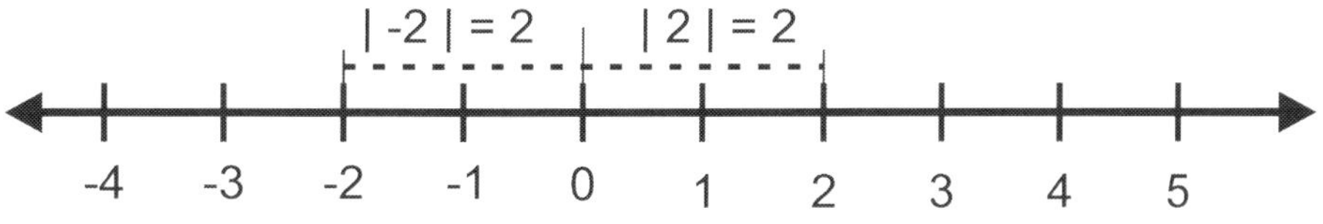

Opposite integers :

The opposite of an integer is the number that is at the same distance from zero in the opposite direction. Every integer has an opposite value, but the opposite of zero is itself.

The opposite of -4 is 4 because it is located the same distance from zero as 4 is , but in opposite direction.

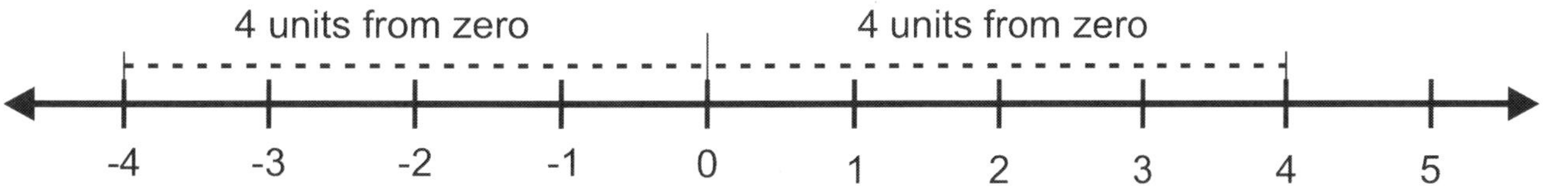

Adding integers using a number line :

The number line is visual representation to understand the addition of positive and negative numbers. Start with the one value on the number line, then add the second value. If the second value (that is added) is positive, we move to the right that many spaces.

If the second value (that is added) is negative, we move to the left that many spaces.
The value where we land on the number line is the solution for the addition of two integers.

Example 1 : (-4) + (5) = 1
Start at the first number, -4, and travel 5 units to the right.

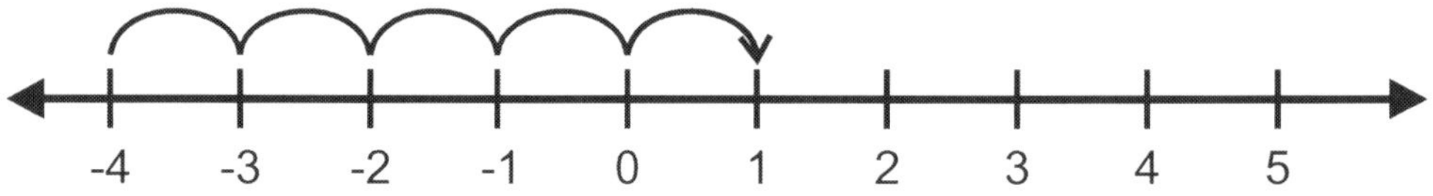

Example 2 : (5) + (-7) = -2
Start at the first number, 5, and travel 7 units to the left.

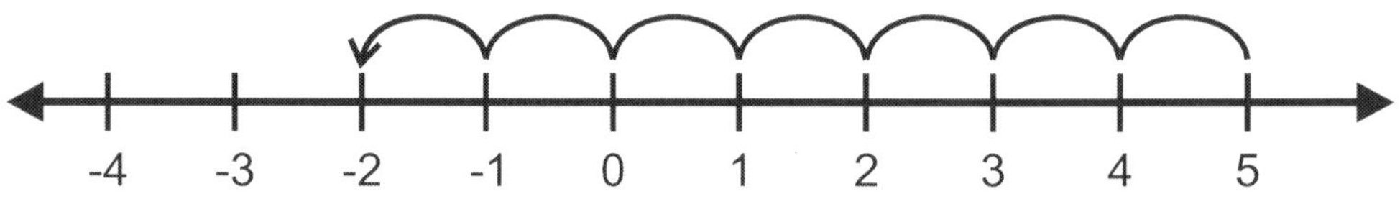

Adding integers using the rules :

Rules for adding integers :

If the signs are the same, add their absolute values, and keep the common sign.

If the signs are different, find the difference between the absolute values of the two numbers, and keep the sign of the number with the greater numerical value.

To the Tune of "Row Your Boat"

Same signs add and keep
Different signs subtract
Keep the sign of the greater digits
then you'll be exact

Subtacting integers using a number line :

A number line is helpful in understanding subtraction of positive and negative values. Start with the first value on the number line, then subtract the second value. If the second value (that is subtracted) is positive, we move to the left that many spaces.

If the second value (that is subtracted) is negative, we move to the right that many spaces. This is because subtraction a negative is the same as adding.
The value where we end on the number line is the answer.

Example 1 : (-2) + (5) = 3
Start at the first number, -2, and travel 5 units to the right.

Subtacting integers using the rules :

Every subtraction problem can be written as an additional problem. When we subtract two integers, just ADD THE OPPOSITE. Subtracting a positive is the same as adding a negative. Subtracting a negative is the same as adding a positive.

Multiplying Integers :

Multiplying integers is same as multiplying whole numbers, except we must keep track of the signs associated to the numbers.

To multiply signed integers, always multiply the absolute values and use these rules to determine the sign of the product value

When we multiply two integers with the same signs, the result is always a positive value.

Positive number X Positive number = Positive number

Negative number X Negative number = Positive number

When we multiply two integers with different signs, the result is always a negative value.

Positive number X Negative number = Negative number

Negative number X Positive number = Negative number

Positive X Positive :	7 X 6 = 42	negative X negative : -7 X -6 = 42
Positive X negative :	7 X -6 = -42	negative X Positive : -7 X 6 = -42

Dividing Integers :

Division of integers is similar to the division of whole numbers, except we must keep track of the signs associated.

To divide signed integers, we must always divide the absolute values and use the below rules to find the quotient value.

When we divide two integers with the same signs, the result is always a positive value.

Positive ÷ Positive = Positive

Negative ÷ Negative = Positive

When we divide two integers with opposite signs, the result is always a negative value.

Positive ÷ Negative = Negative

Negative ÷ Positive = Negative

Examples :

Positive ÷ Positive : 81 ÷ 9 = 9

negative ÷ negative : -81 ÷ -9 = 9

Positive ÷ negative : 81 ÷ -9 = -9

negative ÷ Positive : -81 ÷ 9 = -9

Golden Rules of Integers :

(+) × (+) = (+)
(−) × (−) = (+)
(+) × (−) = (−)
(−) × (+) = (−)

Multiplying or dividing two numbers with same signs will always give a positive value.

Multiplying or dividing two numbers with opposite signs will always give a negative value.

(+) ÷ (+) = (+)
(−) ÷ (−) = (+)
(+) ÷ (−) = (−)
(−) ÷ (+) = (−)

Introduction

Basic rules of number system involves addition, subtraction , multiplication and division.
As the topic of "Numbers" involves some more useful concepts like LCM and GCD,
we shall study them in this chapter. For finding LCM and GCD, the divisibility rules are useful in one way or the other. Hence let us start the learning the divisibility rules.

Divisibility:

In general, if two natural numbers a and b are such that, when 'a' is divided by 'b', a remainder of zero is obtained, we say that 'a' is divisible by 'b'.

For example, 12 is divisible by 3 because 12 when divided by 3, the remainder is zero.

Also, we say that 12 is not divisible by 5, because 12 when divided by 5, it leaves a remainder 2.

Tests of Divisibility:

We now study the methods to test the divisibility of natural numbers with 2, 3, 4, 5, 6, 8, 9 and 11 without performing actual division.

Test of Divisibility by 2:

A natural number is divisible by 2, if its units digit is divisible by 2, i.e.,
the units place is either 0 or 2 or 4 or 6 or 8.

Examples : The numbers 4096, 23548 and 34052 are divisible by '2' as they end with 6, 8 and 2 respectively.

Test of Divisibility by 3:

A natural is divisible by 3 if the sum of its digits is divisible by 3.

Example: Consider the number 2143251. The sum of the digits of 2143251 (2 + 1 + 4 + 3 + 2 + 5 + 1) is 18.

As 18 is divisible by 3, the number 2143251, is divisible by 3.

Test of Divisibility by 4:

A natural number is divisible by 4, if the number formed by its last two digits is divisible by 4.

Examples : 4096, 53216, 548 and 4000 are all divisible by 4 as the numbers formed by taking the last two digits in each case is divisible by 4.

Test of Divisibility by 5:

A natural number is divisible by 5, if its units digit is either 0 or 5.

Examples : The numbers 4095 and 235060 are divisible by 5 as they have in their units place 5 and 0 respectively.

Test of Divisibility by 6:

A number is divisible by 6, if it is divisible by both 2 and 3.

Examples : Consider the number 753618

Since its units digit is 8, so it is divisible by 2. Also its sum of digits = 7 + 5 + 3 + 6 + 1 + 8 = 30, As 30 is divisible by 3, so 753618 is divisible by 3.
Hence 753618 is divisible by 6.

Test of Divisibility by 8:

A number is divisible by 8, if the number formed by its last three digits is divisible by 8.

Examples : 15840, 5432 and 7096 are all divisible by 8 as the numbers formed by last three digits in each case is divisible by 8.

Test of Divisibility by 9:

A natural number is divisible by 9, if the sum of its digits is divisible by 9.

Examples :

(i) Consider the number 125847.
Sum of digits = 1 + 2 + 5 + 8 + 4 + 7 = 27. As 27 is divisible by 9, the number 125847 is divisible by 9.

(ii) Consider the number 145862.
Sum of digits = 1 + 4 + 5 + 8 + 6 + 2 = 26. As 26 not divisible by 9, the number 145862 is not divisible by 9.

Test of Divisibility by 11:

A number is divisible by 11, if the difference between the sum of the digits in odd places and sum of the digits in even places of the number is either 0 or a multiple of 11.

Examples :

(i) Consider the number 9582540
Now (sum of digits in odd places) - (sum of digits in even places)
$= (9 + 8 + 5 + 0) - (5 + 2 + 4)$
$= 11$, which is divisible by 11.
Hence 958254 is divisible by 11.

(ii) Consider the number 1453625
Now, (sum of digits at odd places) - (sum of digits at even places)
$= (1 + 5 + 6 + 5) - (4 + 3 + 2)$
$= 17 - 9 = 8$, which is not divisible by 11

Some Additional Results :

If a natural number N is divisible by two natural numbers a and b, then N is divisible by the product of a and b, if and only if a and b are co-primes.

Examples :

(i) 345 is divisible by 3 as well as by 5, as 3 and 5 are co-primes, 345 is divisible by 15.
(ii) 120 is divisible by 8 and 10, but it is not divisible by 80.

Factors and Multiples :

Lets learn the concepts of factors and multiples.

Definition :

If 'b' divides 'a' leaving a zero remainder, then 'b' is called a factor or divisor of 'a' and 'a' is called the multiple of 'b'.
For example, 6 = 2 3
Here 2 and 3 are factors of 6 (or) 2 and 3 are divisors of 6.
And, 6 is a multiple of 3, 6 is a multiple of 2.

Examples : (i) The factors of 24 = {1, 2, 3, 4, 6, 8, 12, 24}
(ii) The factors of 256 = {1, 2, 4, 8, 16, 32, 64, 128, 256}

Observations :

One is the factor of every natural number and it is the least of the factors of any natural number.
Every natural number is the factor of itself and it is the greatest.

Some Additional Results :

Unique Prime Factorisation Theorem : "Any natural number greater than 1 can be divided into a prime number or a composite number."

For example:

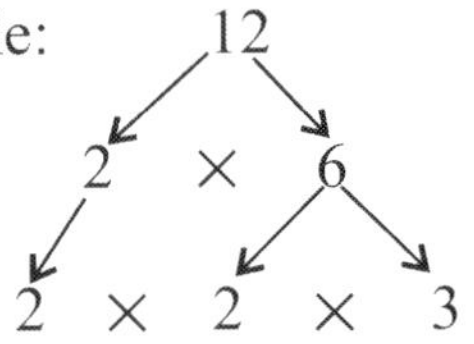

6 is a composite number

All factors are prime numbers.

If a is a composite number of the form $a = b^p c^q d^r$ ..... where b, c, d ... are district prime factors, then the number of factors of $a = (p+1)(q+1)(r+1)$ ....

If a is a composite number of the form $a = b^p c^q d^r$ .... where b, c, d .... are district prime factors, then the sum of all the factors of $a = \frac{(b^{p-1} - 1)}{(b - 1)} \times \frac{(c^{q-1} - 1)}{(c - 1)} \times \frac{(d^{r-1} - 1)}{(d - 1)}$ ......

Perfect Numbers :

A number for which sum of all its factors is twice the number itself is called a perfect number.

Observation :

Euler proved that if $2^k - 1$ is a prime number, then $2^{k-1}(2^k - 1)$ is a perfect number.
A perfect number can never be a prime number.

Examples :

(i) Consider the composite number 6.
Factors of 6 = {1, 2, 3, 6}
Sum of factors = (1 + 2 + 3 + 6) = 12
Clearly, the sum of the factors of 6 is twice the number itself.
Hence 6 is a perfect number.

Example :

(ii) Consider the composite number 48
Factors of 48 = {1, 2, 3, 4, 6, 8, 12, 16, 24, 48}
Sum of factors = 1 + 2 + 3 + 4 + 6 + 8 + 12 + 16 + 24 + 48
= 124 2 48
Clearly, 48 is not a perfect number.

Greatest Common Divisor [GCD] (or) Greatest Common Factor [GCF] (or) Highest Common Factor [HCF]

Definition :

"The greatest common factor of two or more natural numbers is the largest factor in the set of common factors of those numbers." In other words, the GCD (or) GCF of two or more numbers is the largest number that divides each of them exactly.

Example : Find the GCF of 72 and 60.

Solution : Let the set of factors of 72 be A.
A = {1, 2, 3, 4, 6, 8, 9, 12, 18, 24, 36, 72}
Let the set of factors of 60 be B.
B = {1, 2, 3, 4, 5, 6, 10, 12, 15, 20, 30, 60}
The set of common factors for 72 and 60 is A B = {1, 2, 3, 4, 6, 12}
The greatest element in this set is 12
The GCF (or) GCD for 72 and 60 is 2.

Observations :
If two numbers have no factors in common, then their GCF is unity.
i.e., GCF of prime numbers and co-prime numbers is unity.

Methods of finding GCF :

Factors Method :

When the numbers whose GCF has to be found are relatively small, this is the best suited method. Here we resolve the given numbers into their prime factors and find out the largest factor in the set of common factors to given numbers.

This method can be easily applied to any number of numbers.

Examples : (i) Find the GCF of 24 and 36.

Solution :

Resolving given numbers into product of prime factors, we have
36 = ② × ② × ③ × 3
24 = 2 × ② × ② × ③
The common factors to both the numbers are circled.
Now GCF = product common factors of given numbers = 2 × 2 × 3 = 12
GCF (24, 36) = 12

(ii) Find the GCF of 12, 18 and 24.

Solution :

Resolving given numbers into product of prime factors;

12 = ② × 2 × ③
18 = ② × 3 × ③
24 = ② × 2 × 2 × ③

GCF = product of common factors of 12, 18 and 24
= 2 × 3 = 6
GCF = 6

Division Method :

When the numbers whose GCF has to be found are very large, it is time consuming to write down sets of common factors to given numbers;

In this case, we use the method of Long Division. This method was proposed by Euclid and the following steps are involved in it.

Step 1 : Divide the larger number by the smaller number. If the remainder is zero, the divisor is the GCF, otherwise not.

Step 2 : Let the divisor in step 2 be the dividend now, and the remainder of step 1 become the divisor of step 2. Again, if the remainder is zero, the divisor is GCD. Otherwise, step 2 has to be repeated.

Example : Find the GCF of 64 and 56

Solution :
64 divided by 56, quotient is 1 and remainder is 8. Because the reminder is not zero; 56 is not the GCD.

Proceeding further, as mentioned in step 2, 56 is dividend and 8 is divisor. The quotient is 3 and remainder is zero.

Because the remainder is zero, the divisor 8 is the GCD.

GCF of three numbers using division method:

The GCF of 3 numbers is found out by finding GCF of any 2 numbers and GCF of the remaining number with the GCF obtained above.
i.e., GCF (a, b, c) = GCF [GCF (a, b), c].

This process can be extended to any number of numbers.

Example : Find the GCF of 25, 45 and 75.

Solution : Let us first find the GCF of 25 and 45

```
25) 45 (1
    25
    20) 25 (1
        20
         5) 20 (4
            20
             0
```

0

GCF (25, 45) = 5
The GCF of (5, 75) is 5.
The GCF of 25, 45 and 75 is 5.

Observations :

The process of dividing the divisor with quotient is to be repeated until remainder as zero is obtained

If the zero remainder is obtained when the divisor is 1, then the GCD is '1'.

GCD is '1', means that two numbers are relatively prime or co-prime; because they do not have any factor in common other than 1. For eg; 12 and 13 are co-primes.

Some Additional Results :

The largest number which divides p, q and r to give remainders of s, t and u respectively will be the GCD of the three numbers (p – s), (q – t) and (r – u).

The largest number which divides the numbers p, q and r and gives the same remainder in each case will be the GCD of the differences of two or the three numbers (p – q, q – r, p – r).

Least Common Multiple [LCM] :

Definition :

"The least common multiple of two or more natural numbers is the least of their common multiples". In other words, the LCM of two or more numbers is the least number which can be divided exactly by each of the given numbers.

Note : If the set of common multiples is denoted by C, then N and the number of elements in C is infinite and the least element in C is their LCM.

Example : Find L.C.M. of 24 and 36.

Solution : Resolving 24 and 36 into product of prime factors
24 = ② × ② × 2 × ③
36 = ② × ② × ③ × 3

The common prime factors of 24 and 36 are 2, 2 and 3. (which are circled)

The remaining prime factors of 24 is 2. (which is not circled).

The remaining prime factors of 36 is 3. (which is not circled).

LCM = Common factors of 24 the prime factors left in 24 the prime factors left in 36
$= 2 \times 2 \times 3 \times 2 \times 3$
$= 72$

Methods of finding LCM :

Factors Method :

Here the given numbers are decomposed into product of prime factors; from which, the least common multiple is found by multiplying the terms containing factors of numbers raised to their highest powers.

Example : Find the LCM of 32 and 24.

Solution : Resolving given numbers into product of common factors, we have
$32 = 2^5$; $24 = 2^3 \times 3$
LCM = product of terms containing highest powers of factors 2, 3
$= 2^5\ 3 = 96$

LCM of three numbers using factors method:

The method above can be extended in a similar way to three numbers. This is illustrated below:

Example : Find the LCM of 12, 48 and 36

Solution : Resolving given numbers into product of common factors, we have
$12 = 2^2 \times 3^1$; $48 = 2^4 \times 3^1$; $36 = 2^2 \times 3^2$. Then their LCM $= 2^2 \times 3^2 = 16 \times 9 = 144$

Synthetic Division Method of Finding LCM:

LCM of numbers can also be found using synthetic division method. This is illustrated below:

Example : Find the LCM of 144 and 156.

Solution : Using synthetic division, we have:

2	144, 156
2	72, 78
3	36, 39
	12, 13

LCM = 2 2 3 12 13
= 1872

Examples : Find the LCM of 12, 18 and 24

Solution : Using synthetic division;

2	12, 18, 24
2	6, 9, 12
2	3, 9, 6
	1, 3, 2

LCM = 2 2 2 1 3 2 = 48

Some Additional Results :

Any natural N number which when divided by p, q or r leaving the same remainder s in each case will be of the form N = K (LCM of p, q, r) + s, where K = 0, 1, 2, 3 ….

Any natural number N which when divided by p, q and r leaves respective remainders of s, t and u where (p – s) = (q – t) = (r – u) = v (say), then it will be of the form
N = k (LCM of p, q and r) – v, where k = 1, 2, 3 ….

A natural N number which when divided by p and q leaves remainders r and s, will be of the form N = k (LCM of p and q) + n when n is the smallest integer solution for the equations $n = pm_1 + r$ and $n = qm_2 + s$, where m_1 and m_2 are natural numbers.

Relationship between LCM and GCF :

The LCM and GCF of two given numbers are related to the given numbers by the following relationship.
Product of the numbers = LCM × GCF

where, LCM denotes the LCM of the given numbers and GCF denotes the GCF of the given numbers.

Example : Consider two numbers, 24 and 36.

These can be resolved into product of prime factors as below :

$24 = 2^3 \times 3$

$36 = 2^2 \times 3^2$

Now LCM $(24, 36) = 2^3 \times 3^2 = 72$

GCF $(24, 36) = 2^2 \times 3 = 12$

Now; Product of numbers $= 24 \times 36 = 2^5 \times 3^3 = 864$

Product of LCM and GCF $= 72 \times 12 = 2^5 \times 3^3 = 864$

Clearly, Product of the numbers = Product of the LCM and GCF.

Relatively Prime Numbers :

Definition :

If two numbers do not have any common factors other than 1, then they are called relatively prime numbers or co-prime numbers.

Concept : We know, every number has at least two factors, 1 and itself. If it has more than two factors, it is a composite number and if it does not have any factor except 1 and itself, it is a prime number.

But if two numbers (prime or composite) are such that they have only one common factor '1' are called relatively prime.

Example : Consider three numbers 8, 18 and 25

Now,

A, The set of factors of 8 = {1, 2, 4, 8}

B, The set of factors of 18 = {1, 2, 3, 6, 9, 18}

C, The set of factors of 25 = { 1, 5, 25}

Now, A B = {1, 2} ; B C = {1} ; C A = {1}

Clearly the common factors for both (18, 25) and (8, 25) is 1 only.

They are generally written as (18, 25) = 1 and (8, 25) = 1

Note : Here; 18, 25 and 8 are not prime numbers (composite) but they are relatively prime numbers.

Observations :

The G.C.D. of two relatively prime numbers is 1 and their LCM is product of the numbers

Any two prime numbers are always relatively prime to each other.

Two relatively prime numbers need not be prime numbers.

LCM and GCF of Fractions :

The LCM and GCF of fractions can be determined by the following relations :

$$\text{LCM of fractions} = \frac{\text{LCM of numerators}}{\text{GCD of denominators}}$$

$$\text{GCF of fractions} = \frac{\text{GCD of numerators}}{\text{LCM of denominators}}$$

Examples : Find the GCD and LCM of $\frac{4}{5}$, $\frac{2}{5}$ and $\frac{3}{4}$

Solutions : $\text{LCM}\left(\frac{4}{5}, \frac{2}{5}, \frac{3}{4}\right) = \frac{\text{LCM }(4, 2, 3)}{\text{HCF }(5, 5, 4)} = \frac{4 \times 3}{5}$

$$\text{GCD }\left(\frac{4}{5}, \frac{2}{5}, \frac{3}{4}\right) = \frac{\text{HCF }(4, 2, 3)}{\text{LCM }(5, 5, 4)} = \frac{1}{5 \times 4} = \frac{1}{20}$$

Introduction of Exponents

Whenever any integer, let us say x, is added n times, the result obtained will be equal to n times x i.e. nx. But in case, if the integer x is multiplied for n times, the result obtained will be equal to x^n (which is called as exponential form). The problems relating to these will be studied under 'Exponents'. We shall look at the rules/properties pertaining to these exponential numbers in this chapter.

Rational Exponents and Radicals

If 'a' is any real number and 'n' is a positive integer, then the product $a \times a \times a \times$ ---- n times is represented by the notation a^n. This notation is referred to as exponential form. In the above notation, a is called the base and n is called the power or exponent or index (plural of index is indices). a^n is read as 'n^{th} power of a' or 'a to the power n'.

Example: $6 \times 6 \times 6 \times 6 \times 6 \times 6 \times 6$ can be written as 6^7. Here 6 is called as base and 7 is called as index (or exponent).

For a non-zero rational number 'a' with a negative integral exponent 'm' the following result can be observed.

$a^m = a^{-n} = a^{-1} \times a^{-1} \times a^{-1} \times a^{-1} \times$ - - - - $\times$ - - - n times

$= \frac{1}{a} \times \frac{1}{a} \times \frac{1}{a} \times \frac{1}{a}$ - - - - n times $= \left(\frac{1}{a}\right)^n = \frac{1}{a^n}$

Example: $6^{3} = \left(\frac{1}{6}\right)^3$

Rational Indices:

1. n^{th} root of a :

A real number x is said to be the nth root of a if $x^n = a$; where a is any real number and n is a positive integer.

Parts of the exponent:

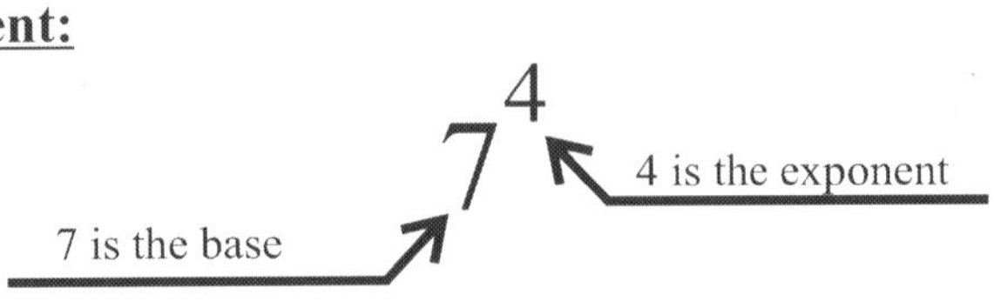

This is read as "Seven to the fourth power"

The nth root of a can be represented as $a^{1/n}$ or $\sqrt[n]{a}$. Here $a^{1/n}$ is called exponential form and the form $\sqrt[n]{a}$ is called radical form. The sign $\sqrt[n]{\ }$ is called radical sign and $\sqrt[n]{a}$ is called radical. The number n is a positive integer is called the index of radical and a is called the radicand.

Example: We know that $32 = 2^5$. So we can say that 2 is 5th root of 32. It is written as $32^{1/5} = 2$ or $\sqrt[5]{32} = 2$
Similarly $\sqrt[3]{64} = 3$, $\sqrt[4]{625} = 5$, $\sqrt[6]{64} = 2$, etc.

Note:

(i) If n is negative as in case $64^{\frac{1}{3}}$, we write the radical form as follows;

$$64^{\frac{1}{3}} = \left(\frac{1}{64}\right)^{\frac{1}{3}} = \sqrt[3]{\frac{1}{64}}.$$

The radical form of $64^{\frac{1}{3}}$ should not be taken as $\sqrt[3]{64}$ as in the radical form $\sqrt[n]{a}$ of n is a positive integer.

i.e., $a^{1/n} = \sqrt[n]{\frac{1}{a}}$, where n is a positive integer.

(ii) $\sqrt[n]{a}$ is positive for $a > 0$ and n being a positive integer.
Example: $\sqrt[5]{32} = 2$, $\sqrt[6]{64} = 2$, $\sqrt[3]{27} = 3$, etc.

(iii) $\sqrt[n]{a}$ is negative for $a < 0$ and n being any odd positive integer.
Example: $\sqrt[3]{8} = -2$, $\sqrt[7]{128} = -2$, $\sqrt[5]{243} = -3$, etc.

(iv) $\sqrt[n]{a}$ does not exist in set of real numbers, for $a < 0$ and n being even positive integer.
Example: $\sqrt[2]{16}$, $\sqrt[4]{256}$, $\sqrt[6]{64}$ etc doesn't exist.

Each positive number has two square roots, one positive and the other negative.

Example: $\sqrt{36} = 6$ or -6 (since $6^2 = (-6)^2 = 36$).

If a is a positive rational number and n = p/q is a positive rational exponent, then we can define $a^{p/q}$ in two ways.

(1) $a^{\frac{p}{q}}$ is the qth root of a^p, i.e. $a^{\frac{p}{q}} = \left(a^p\right)^{\frac{1}{q}}$

(2) $a^{\frac{p}{q}}$ is the pth power of qth root of a, i.e. $a^{\frac{p}{q}} = \left(a^{\frac{1}{q}}\right)^p$.

Laws of indices:

For all real numbers a and b and all rational numbers m and n, we have

(i) $a^m \times a^n = a^{m+n}$

Examples: (1) $2^3 \times 2^6 = 2^{3+6} = 2^9$

(2) $\left(\frac{5}{6}\right)^4 \times \left(\frac{5}{6}\right)^5 = \left(\frac{5}{6}\right)^{4+5} = \left(\frac{5}{6}\right)^9$

(3) $5^{2/3} \times 5^{4/3} = 5^{(2/3+4/3)} = 5^{6/3} = 5^2$

(4) $2^3 \times 2^4 \times 2^5 \times 2^8 = 2^{(3+4+5+8)} = 2^{20}$.

(5) $\left(\sqrt{7}\right)^3 \times \left(\sqrt{7}\right)^{\frac{5}{2}} = \left(\sqrt{7}\right)^{3+\frac{5}{2}} = \left(\sqrt{7}\right)^{\frac{11}{2}}$

(ii) $a^m \div a^n = a^{m-n}, a \neq 0$

Examples: (a) $7^8 \div 7^3 = 7^{8-3} = 7^5$

(b) $\left(\frac{7}{3}\right)^9 \div \left(\frac{7}{3}\right)^5 = \left(\frac{7}{3}\right)^{9\ \ 5} = \left(\frac{7}{3}\right)^4$

(c) $9^{\frac{2}{3}} \div 9^{\frac{1}{6}} = 9^{\left(\frac{2}{3}-\frac{1}{6}\right)} = 9^{\left(\frac{4-1}{6}\right)} = 9^{\frac{3}{6}} = 9^{\frac{1}{2}}$

(d) $\left(\frac{5}{7}\right)^{\frac{8}{9}} \div \left(\frac{5}{7}\right)^{\frac{1}{3}} = \left(\frac{5}{7}\right)^{\left(\frac{8}{9}-\frac{1}{3}\right)} = \left(\frac{5}{7}\right)^{\frac{8-3}{9}} = \left(\frac{5}{7}\right)^{\frac{5}{9}}$

Note: $a^n \div a^n = 1$
or $a^{n-n} = a^0 = 1$
$\therefore a^0 = 1, a \neq 0$

(iii) $(a^m)^n = a^{m \times n}$

Examples: (a) $(5^2)^3 = 5^{2 \times 3} = 5^6$

(b) $$\left[\left(\frac{2}{3}\right)^4\right]^5 = \left(\frac{2}{3}\right)^{4\times5} = \left(\frac{2}{3}\right)^{20}$$

(c) $$\left[\left(\frac{5}{7}\right)^{\frac{2}{3}}\right]^{\frac{9}{8}} = \left(\frac{5}{7}\right)^{\left(\frac{2}{3}\times\frac{9}{8}\right)} = \left(\frac{5}{7}\right)^{\frac{3}{4}}$$

(iv) $$\left(\frac{a}{b}\right)^n = \frac{a^n}{b^n}$$

Example: $$\left(\frac{4}{5}\right)^7 = \frac{4^7}{5^7}$$

Note: Conversely we can write $\left(\frac{a^n}{b^n}\right) = \left(\frac{a}{b}\right)^n$

Example: $$\frac{8}{27} = \frac{2^3}{3^3} = \left(\frac{2}{3}\right)^3$$

(v) $(ab)^n = a^n \times b^n$

Examples: (a) $20)^5 = (4 \times 5)^5 = 4^5 \times 5^5$
(b) $(42)^7 = (2 \times 3 \times 7)^7 = 2^7 \times 3^7 \times 7^7$

Note: Conversely we can write $a^n \times b^n = (ab)^n$

Examples: (a) $4^8 \times 5^8 = (4 \times 5)^8 = 20^8$

(b) $\left(\frac{2}{3}\right)^5 \times \left(\frac{9}{8}\right)^5 = \left(\frac{2}{3} \times \frac{9}{8}\right)^5 = \left(\frac{3}{4}\right)^5$

(vi) $a^{-n} = \frac{1}{a^n}$, $a \neq 0$

Example: $2^{-4} = \frac{1}{2^4}$, $5^{-1} = \frac{1}{5}$

Note: $a^{-1} = \frac{1}{a^1} = \frac{1}{a}$

(vii) $\left(\frac{a}{b}\right)^n = \left(\frac{b}{a}\right)^n$

Examples: (a) $\left(\frac{5}{9}\right)^3 = \left(\frac{9}{5}\right)^{-3}$

(b) $\left(\frac{1}{5}\right)^{-1} = \left(\frac{5}{1}\right)^1 = 5$

Note: $\left(\frac{1}{a}\right)^{-1} = \left(\frac{a}{1}\right)^1 = a$

(viii) If $a^m = a^n$, then $m = n$, where $a \neq 0$, $a \neq 1$

Examples: (a) If $5^p = 5^3 \Rightarrow p = 3$

(b) If $4^p = 256$

$4^p = 4^4 \Rightarrow p = 4$

(ix) For positive numbers a and b, if $a^n = b^n$, $n \neq 0$, then $a = b$ (when n is odd)

Examples: (a) If $5^7 = p^7$, then clearly $p = 5$.

(b) If $(5)^{2n-1} = (3 \times p)^{2n-1}$, then clearly $5 = 3p$ or $p = 5/3$

(x) If $p^m \times q^n \times r^s = p^a\, q^b\, r^c$, then m =a, n = b, s = c, where p, q, r are different primes.

Examples: (a) If $40500 = 2^a \times 5^b \times 3^c$, then find $a^a \times b^b \times c^c$

2	40,500
2	20,250
5	10,125
5	2,025
5	405
3	81
3	27
3	9
	3

$\therefore 40500 = 2^2 \times 5^3 \times 3^4 = 2^a \times 5^b \times 3^c$

$\therefore a = 2, b = 3, c = 4$, [Using the above law].

$\therefore a^a \times b^b \times c^c = 2^2 \times 3^3 \times 4^4 = 27{,}648$

Example 8 : (a) $20)^5 = (4 \times 5)^5 = 4^5 \times 5^5$

(b) $(42)^7 = (2 \times 3 \times 7)^7 = 2^7 \times 3^7 \times 7^7$

Note: Conversely we can write $a^n \times b^n = (ab)^n$

Example 9 : (a) $((5)^3)^2 = 5^{3X2} = (5)^6 = 5\times5\times5\times5\times5\times5 = 15625$

(b) $(2)^3 \times (2)^5 = (2)^{3+5} = (2)^8 = 2\times2\times2\times2\times2\times2\times2\times2 = 256$

(c) $(7)^0 = 1$ Any base value rise to the power zero is always equal to 1

(d) $(3)^{-4} = \dfrac{1}{(3)^4} = \dfrac{1}{3\times3\times3\times3} = \dfrac{1}{81}$

(d) $\dfrac{(8)^7}{(8)^5} = (8)^{7-5} = (8)^2 = 64$

(e) $\dfrac{(9)^4}{(9)^7} = (9)^{4-7} = (9)^{-3} = \dfrac{1}{(9)^3} = \dfrac{1}{9\times9\times9} = \dfrac{1}{729}$

(f) $(2)^4 = 2\times2\times2\times2 = 16$

(g) $(-2)^4 = -2\times-2\times-2\times-2 = 16$

(h) $-(2)^4 = -(2\times2\times2\times2) = -16$

(i) $-(2)^3 = -(2\times2\times2) = -8$

(j) $(-2)^3 = -2\times-2\times-2 = -8$

(k) $-(-2)^3 = -(-2\times-2\times-2) = -(-8) = 8$

Tip 1 : When the exponent is an even number the simplified value is always positive, when the base has a positive or negative value.

Tip 2 : When the exponent is an odd number the simplified value is always positive, when the base has a positive value.

Tip 3 : When the exponent is an odd number the simplified value is always negative, when the base has a negative value.

The percentage symbol is a representation of percentage. In statistics, percentages are often left in their base form of 0 - 1, where 1 represents the whole. We multiply the decimal by a factor of 100 to find the percentage.

Percent Equation form :

Percentages can be setup as proportions. The parts of a percent can be found by setting up a proportion as below.

$$\frac{\text{is}}{\text{of}} = \frac{\text{percent}}{100} \quad \text{OR} \quad \frac{\text{part}}{\text{total}} = \frac{\text{percent}}{100}$$

Before we calculate a percentage, we should understand exactly what a percentage is ? The word percentage comes from the word percent. If you split the word percent into its root words, you see "per" and "cent." Cent is an old European word with French, Latin, and Italian origins meaning "hundred". So, percent is translated directly to "per hundred." If we have 39 percent, we literally have 39 per 100. If it snowed 16 times in the last 100 days, it snowed 16 percent of the time.

Whole numbers, decimals and fractions are converted to percentages. Decimal format is easier to convert into a percentage. Fractions can be converted into decimals and then to percentages.

Converting a decimal to a percentage is as simple as multiplying it by 100. To convert 0.92 to a percent, simply multiple 0.92 by 100.

$0.92 \times 100 = 92\%$

A percentage is an expression of part of the whole. Nothing is represented by 0%, and the whole amount is 100%. Everything else is somewhere in between 0 and 100.

For example, say you have 20 muffins. If you share 12 muffins with your friends, then you have shared 12 out of the 20 muffins ($\frac{12}{20} \times 100\% = 60\%$ shared). If 20 muffins are 100% and you share 60% of them, then $100\% - 60\% = 40\%$ of the muffins are still left.

Determine the part of the whole :

Given the value for part of the whole and the whole.
OR
Given two parts that make up the whole.

It is important to differentiate what the percentage is "of."
Example : A jar contains 99 gold beads and 51 blue beads. A total of 150 beads are in the jar. In this example, 150 beads makes up a whole jar of beads, i.e. 100%._

1. **Put the two values into a fraction.** The part goes on top of the fraction (numerator), and the whole goes on the bottom (denominator). Therefore the fraction in this case is $\frac{99}{150}$ (part/whole) or $\frac{51}{150}$ (part/whole).

2. **Convert the fraction into a decimal.** Convert $\frac{99}{150}$ into decimal, divide 99 by 150

$$\frac{99}{150} = 0.66$$

Converting the decimal into a percent :

Multiply the result obtained in the step above by 100% (per 100 = *per cent*).
0.66 multiplied by 100 equals 66%. Gold beads are 66% of the total beads in the jar.

Converting the percentage into a decimal : Working backward from before, divide the percentage by 100, or you can multiply by 0.01.

$$66\% = \frac{66}{100} = 0.66$$

Re-word the problem with your new values :

Given in the form of "**X** of **Y** is **Z**." X is the decimal form of your percent, "of" means to multiply, Y is the whole amount, and Z is the answer.

Example: 3% of $25 is 0.75.
The amount of interest accrued each day on a 3% of loan amount $25 s 0.75

Discounts :

Discounts are offered on the original price of the item on sale.

1. The discount percent need to be subtracted from the whole, meaning if a discount of 20% is offered on an item priced at \$60 then sale price is 80% of \$60.
   100% - 20% = 80% (whole percent - discount percent).

   80% of \$60 = $\frac{80}{100}$ X 60 = \$48

Markup percentage :

Markup percentage is also referred as gain percentage. Markups are added to the original price of the item before selling it. The greater the markup percentage higher are the profits.

1. The markup percent needs to be added to the whole, meaning if a markup of 20% is added on an item priced at \$60 then sale price is 120% of \$60.
   100% + 20% = 120% (whole percent + markup percent).

   120% of \$60 = $\frac{120}{100}$ X 60 = \$72

Tax percentage :

Taxes are added to the original price of the item before selling it to the customer.

1. The tax percent needs to be added to the whole, meaning if a tax of 5% is added on an item priced at \$60 then sale price is 105% of \$60.
   100% + 5% = 105% (whole percent + tax percent).

   105% of \$60 = $\frac{105}{100}$ X 60 = \$63

Simple Interest :

Simple interest is an easy way of calculating interest on a loan amount.

I = PTR

I = Simple Interest
P = Principle amount of loan
T = Term of the loan in years
R = Rate of interest in decimal form.

Remember final loan amount (A) = P + I

Or

$$I = \frac{PTR}{100}$$

I = Simple Interest
P = Principle amount of loan
T = Term of the loan in years
R = Rate of interest in numerical value only (not in decimal form since the formula has 100 in the denominator).

Remember final loan amount (A) = P + I

Example : Dan took a loan of $1,650 at 7% rate of interest for 5 years from his friend Johnson. Using simple interest find the loan amount to be paid to Johnson at the end of the term.

$$I = \frac{PTR}{100}$$

P = $1,650
R = 7
T = 5 years

$$I = \frac{1650 \times 7 \times 5}{100}$$

$$I = \frac{57750}{100}$$

$$= \$577.50$$

Interest to be paid at the end of the loan term is $577.50
Remember the original loan amount $1,650 also need to be paid of.
Amount to be paid of (A) = P + I

A = $1,650 + $577.50
= $2,227.50

Compound interest :

Compound interest is a complex way of calculating interest on a loan amount and the interest on the interest accurred.

$$A = P(1 + \frac{r}{n})^{nt}$$

Where A = Amount at the end of the loan term
Note :It includes principle amount and the interest

P = Principle loan amount
r = Rate of loan or loan percent or interest rate.
Note : Rate needs to be converted to decimal.

n = Number of times interest is compounded per year.
t = Loan term in years.

Remember final loan amount (A) = P + I

OR

$$A = P(1 + \frac{r}{100n})^{nt}$$

Where A = Amount at the end of the loan term
Note :It includes principle amount and the interest

P = Principle loan amount
r = Rate of loan or loan percent or interest rate.
Note : Rate need not be converted to decimal.

n = Number of times interest is compounded per year.
t = Loan term in years.

Remember final loan amount (A) = P + I

Example : Dan took a loan of \$49,100 at 14% rate of interest for 2 years compounded annually from his friend Johnson. Using compound interest find the loan amount to be paid to Johnson at the end of the term.

$$A = P\left(1 + \frac{r}{100n}\right)^{nt}$$

P = \$49,100
R = 14
T = 2 years
n = 1

$$A = 49100\left(1 + \frac{14}{100}\right)^{2}$$

$$A = 49100\,(1.14)^{2}$$

$$A = \$63{,}810.36$$

$$A = P + I \text{ OR } I = A - P$$

Interest I = \$63,810.36 - \$49,100 = \$14,710.36

Interest to be paid at the end of the loan term is \$14,710.36
Remember the original loan amount \$49,100 also need to be paid of.
Amount to be paid of (A) = P + I

$$A = \$63{,}810.36$$

Dan needs to pay \$63,810.36 to Johnson at the end of the loan term.

Example : Dan took a loan of \$49,100 at 14% rate of interest for 2 years compounded semi annually from his friend Johnson. Using compound interest find the loan amount to be paid to Johnson at the end of the term.

$$A = P\left(1 + \frac{r}{100n}\right)^{nt}$$

P = \$49,100
R = 14
T = 2 years
n = 2

$$A = 49100\left(1 + \frac{14}{200}\right)^{4}$$

$$A = 49100\,(1.07)^{4}$$

A = \$64,360.08

A = P + I OR I = A - P

Interest I = \$64,360.08 - \$49,100 = \$15,260.08

Interest to be paid at the end of the loan term is \$15,260.08
Remember the original loan amount \$49,100 also need to be paid of.
Amount to be paid of (A) = P + I

A = \$64,360.08

Dan needs to pay \$64,360.08 to Johnson at the end of the loan term.

InEqualities and Absolute value of a Number:

If x is a real number, then its absolute value or modulus value is denoted by $|x|$ (read as mod x) and defined as follows,

$|x| = x$ when $x > 0$.

$|x| = -x$ when $x < 0$.

$|x| = 0$ when $x = 0$.

For real numbers A and B , an equation as $|A| = B$ when $B = 0$
has 2 solutions $A = B$ or $A = -B$.
If $B < 0$ then $|A| = B$ has no solution.

Set	Notation	Interval Notation
All real numbers between a and b, not including a or b	$\{x \mid a < x < b\}$	(a , b)
All real numbers greater than a, not including a	$\{x \mid x > a\}$	(a , ∞)
All real numbers less than b, not including b	$\{x \mid x < b\}$	$(-\infty , b)$
All real numbers greater than a, including a	$\{x \mid x \geq a\}$	$[a , \infty)$
All real numbers less than b, including b	$\{x \mid x \leq b\}$	$(-\infty , b]$
All real numbers between a and b, including a	$\{x \mid a \leq x < b\}$	$[a , b)$
All real numbers between a and b, including b	$\{x \mid a < x \leq b\}$	$(a, b]$
All real numbers between a and b, including a and b	$\{x \mid a \leq x \leq b\}$	$[a , b]$
All real numbers less than a or greater than b	$\{x \mid x < a \text{ and } x > b\}$	$(-\infty , a) \cup (b , \infty)$
All real numbers	$\{x \mid x \text{ is all real numbers}\}$	$(-\infty , \infty)$

Properties of inequalities :

1. Addition property
   If $a < b$, then $a + c < b + c$

2. Multiplication property
   If $a < b$ and $c > 0$, then $ac < bc$
   If $a < b$ and $c < 0$, then $ac > bc$

Example:

The value of $|7| = 7 \ (7 > 0)$

The value of $|-5| = -(-5) = 5 \, (-5 < 0)$

The value of $|x|$ is always positive.

Range of $|x|$ is positive real numbers including zero.

Properties:

(i) If $|x| = a$ then $x = \pm a$.

(ii) If $|x| \le a$ then $-a \le x \le a$.

(iii) If $|x| \ge a$ then $x \le -a$ or $x \ge a$.

Example 1: If $|x| = 8$, then find the values of x.

Solution: If $|x| = a$; $x = \pm a$

$|x| = 8$, $x = \pm 8$.

Example 2: If $|x| \le 9$, then find the range of x.

Solution: If $|x| \le 9$, then $-a \le x \le a$

$|x| \le 9 \Rightarrow -9 \le x \le 9$.

Example 3: If $|x| \geq 15$, then find the range of x.

Solution: If $|x| \geq a$, the $x \leq -a$ or $x \geq a$.

$|x| \geq 15 \Rightarrow x \leq -15$ or $x \geq 15$.

Example 4: If $|x+5| = 9$, find the value of x.

Solution: $|x| = a$, then $x = \pm a$

$|x+5| = 9 \Rightarrow x + 5 = \pm 9$.

$x + 5 = 9$ or $x + 5 = -9$

$x = 4$ or $x = -14$.

Example 5: $|3x-7| = 22$, find the value of x.

Solution: $|3x-7| = 22 \Rightarrow 3x - 7 = \pm 22$

$3x - 7 = 22$ or $3x - 7 = -22$

$3x = 29$ or $3x = -22 + 7$

$x = \frac{29}{3}$ or $3x = -15 \Rightarrow x = -5$

$x = \frac{29}{3}$ or $x = -5$.

Example 6: Solve $|x-5| < 3$.

Solution: $|x| < a \Rightarrow -a < x < a$

$|x-5| < 3 \Rightarrow -3 < x-5 < 3$.

$\Rightarrow -3 + 5 < x < 3 + 5$

$2 < x < 8$.

Example 7: Solve $|3x-10| > 43$.

Solution: $|x| > a \Rightarrow x < -a$ or $x > a$.

$|3x-10| > 43 \Rightarrow 3x - 10 < -43$ or $3x - 10 > 43$.

$3x < -43 + 10$ or $3x > 43 + 10$

$3x < -33$ or $x > \frac{53}{3}$.

$x < -11$ or $x > \frac{53}{3}$.

Graph of $|x|$:

Since x is any real number, $|x|$ always positive, so graph of $|x|$ belongs to first and second quadrants only.

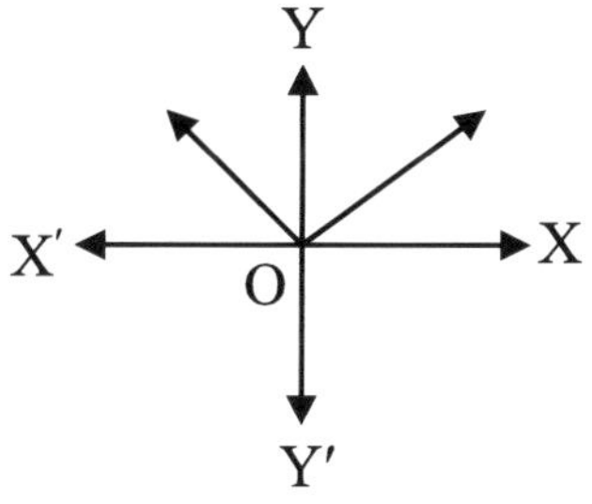

Solving radical equations

An equation containing terms with a variable in the radicand is called a radical equation.

Steps to solve the radical equation :

1. Isolate the radical expression on one side of the equal sign.
2. Move all other terms to other side of the equation.
3. If the radical is a square root , then square both sides of the equation.
   If the radical is a cube root , then rise both sides of the equation to the third power.
   If the radical is a nth root , then rise both sides of the equation to the nth power.
4. If the radical still exists repeat steps 1,2,3. Repeat the process until the radical is eliminated. .
5. Solve the equation.
6. Substitute the solution back for verification.

Solving absolute value equations

The absolute value of x is written as $|x|$. It has the following properties:
If $x \geq 0$, then $|x| = x$.
If $x < 0$, then $|x| = -x$.
For real numbers A and B, an equation of the form $|A| = B$, with $B \geq 0$, will have solutions when $A = B$ or $A = -B$.

If $B < 0$, the equation $|A| = B$ has no solution.
An absolute value equation in the form $|ax + b| = c$ has the following properties:
If $c < 0$, $|ax + b| = c$ has no solution.
If $c = 0$, $|ax + b| = c$ has one solution.
If $c > 0$, $|ax + b| = c$ has two solutions.

1. Isolate the absolute value expression on one side of the equal sign.
2. If $c > 0$, write and solve two equations: $ax+b=c$ and $ax+b=-c$.

The Discriminant

The quadratic formula not only gives the solutions to a quadratic equation ax2 + bx + c = 0, but alsodescribes the nature of the solutions. The discriminant, b2 – 4ac determines whether the solutions are real numbers or complex numbers. Below table describes the properties of the discriminant to the solutions of a quadratic equation.

Value of Discriminant	Results
$b^2 - 4ac = 0$	One rational solution (double solution)
$b^2 - 4ac > 0$, perfect square	Two rational solutions
$b^2 - 4ac > 0$, not a perfect square	Two irrational solutions
$b^2 - 4ac < 0$	Two complex solutions

polynomial equations

A polynomial of degree n is an expression of the type

$$a_n x^n + a_{n-1} x^{n-1} + \cdots + a_2 x^2 + a_1 x + a_0$$

where n is a positive integer and $a_n, \ldots, a_0$ are real numbers and $a_n \neq 0$.

Setting the polynomial equal to zero gives a **polynomial equation**. The total number of solutions (real and complex) to a polynomial equation is equal to the highest exponent n.

Solving Radical Equations:

An equation containing terms with a variable in the radicand is called a **radical equation**.

Given a radical equation, solve it.

1. Isolate the radical expression on one side of the equal sign. Put all remaining terms on the other side.
2. If the radical is a square root, then square both sides of the equation. If It is a cube root , then raise both sides of the equation to the third power. In other words, for an n the root radical, raise both sides to the nth power. Doing so eliminates the radical symbol.
3. Solve the remaining equation.
4. If a radical term still remains, repeat steps1–2.
5. Confirm solutions by substituting the min to the original equation.

Set Indicated	Set-Builder Notation	Interval Notation
All real numbers between a and b, but not including a or b	$\{x \mid a < x < b\}$	(a, b)
All real numbers greater than a, but not including a	$\{x \mid x > a\}$	(a, ∞)
All real numbers less than b, but not including b	$\{x \mid x < b\}$	$(-\infty, b)$
All real numbers greater than a, including a	$\{x \mid x \geq a\}$	$[a, \infty)$

Set Indicated	Set-Builder Notation	Interval Notation
All real numbers less than b, including b	$\{x \mid x \leq b\}$	$(-\infty, b]$
All real numbers between a and b, including a	$\{x \mid a \leq x < b\}$	$[a, b)$
All real numbers between a and b, including b	$\{x \mid a < x \leq b\}$	$(a, b]$
All real numbers between a and b, including a and b	$\{x \mid a \leq x \leq b\}$	$[a, b]$
All real numbers less than a or greater than b	$\{x \mid x < a \text{ and } x > b\}$	$(-\infty, a) \cup (b, \infty)$
All real numbers	$\{x \mid x \text{ is all real numbers}\}$	$(-\infty, \infty)$

properties of inequalities

Addition Property If $a < b$, then $a + c < b + c$.

Multiplication Property If $a < b$ and $c > 0$, then $ac < bc$.

If $a < b$ and $c < 0$, then $ac > bc$.

These properties also apply to $a \leq b$, $a > b$, and $a \geq b$.

Geometry Notes

1. Complementary Angles:

Two angles that add up to 90 degrees are called as Complementary angles.

Example: $\angle X + \angle Y = 90$

$\angle X$ and $\angle Y$ are called complementary angles.

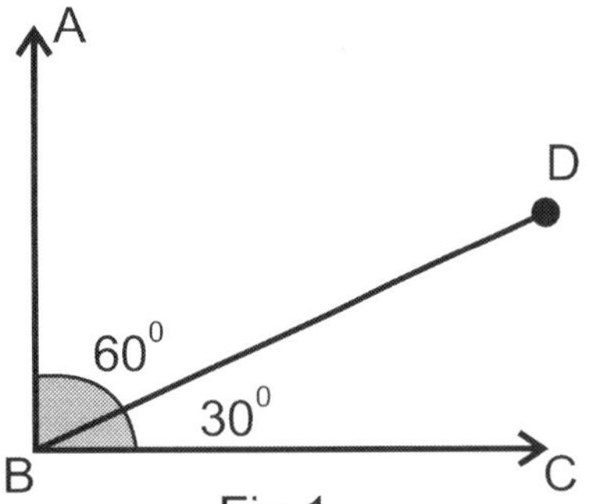

Fig 1

$30^0 + 60^0 = 90^0$

30^0 and 60^0 are complementary angles

2. Supplementary Angles:

Two angles that add up to 180 degrees are called as supplementary angles.

Example: $\angle a + \angle b = 180$

$\angle a$ and $\angle b$ are called complementary angles.

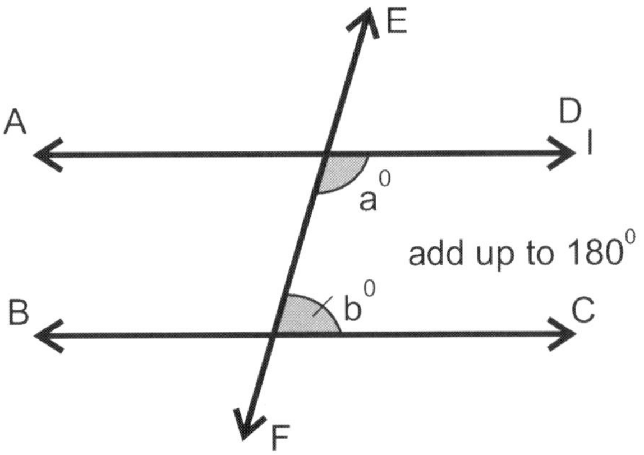

Fig 2

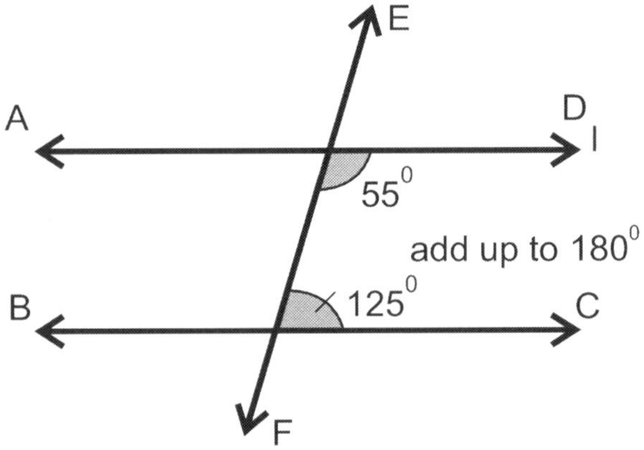

Fig 3

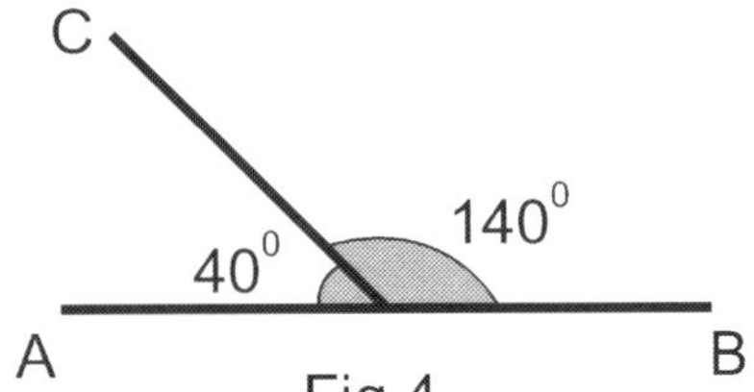

Fig 4

$\angle a + \angle b = 180$; $55^0 + 125^0 = 180^0$

55^0 and 125^0 are called as complementary angles

$\angle a + \angle b = 180$; $40^0 + 140^0 = 180^0$

40^0 and 140^0 are called as complementary angles

3. Vertical Angles:

Vertical angles are pairs of opposite angles made by intersecting lines.
If two angles are vertical, then they are congruent.

Example:

$\angle a$ and $\angle b$ are called vertical angles

$\angle a = \angle b$

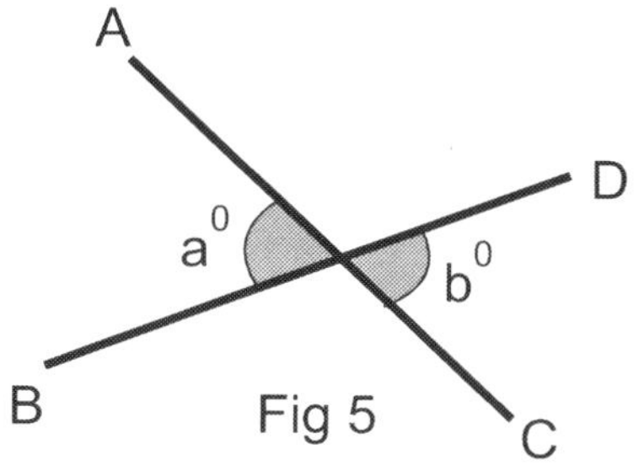

Fig 5

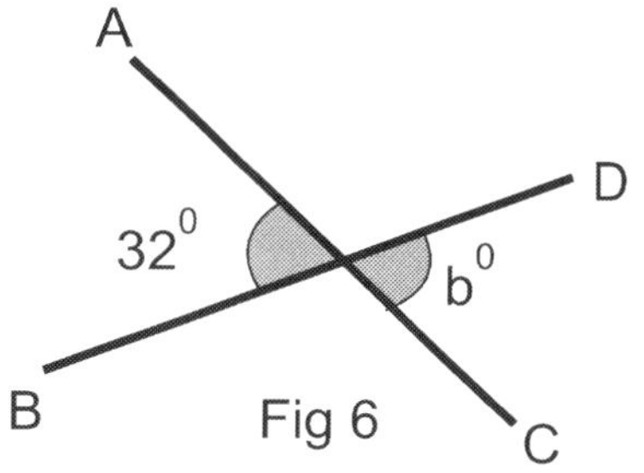

Fig 6

$\angle a$ and $\angle b$ are called vertical angles and vertical angles are equal to each other.
In Fig 6 based on the rule of the vertical angles $\angle b = 32^0$

4. Adjacent Angles:

Two angles that have a common side and a common vertex (corner point), and don't overlap are called adjacent angles

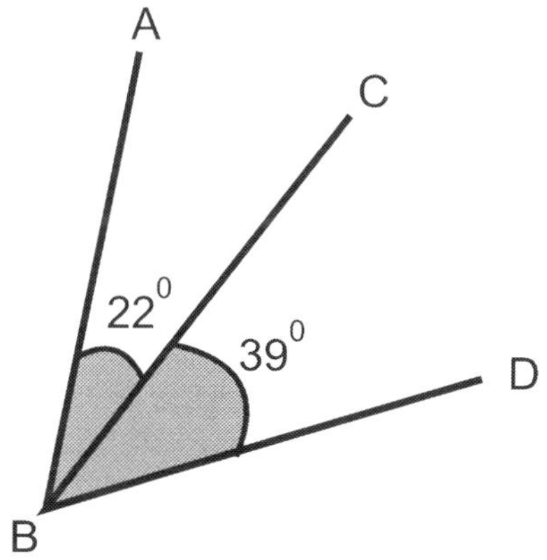

Fig 7

$\angle ABC$ and $\angle CBD$ are called as adjacent angles as they share same vertex B

5. Corresponding Angles

When two lines are crossed by another line (Transversal), the angles in matching corners are called as corresponding angles. A pair of angles each of which is on the same side of one of two lines cut by a transversal and on the same side of the transversal

The angles which occupy the same relative position at each intersection where a straight line crosses two others. If the two lines are parallel, the corresponding angles are equal.

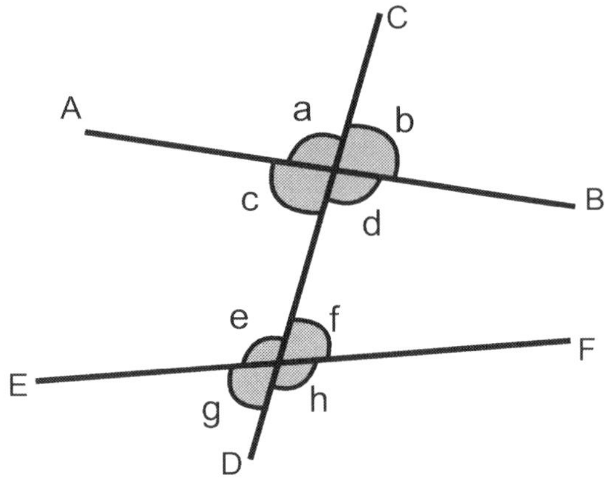

Fig 8

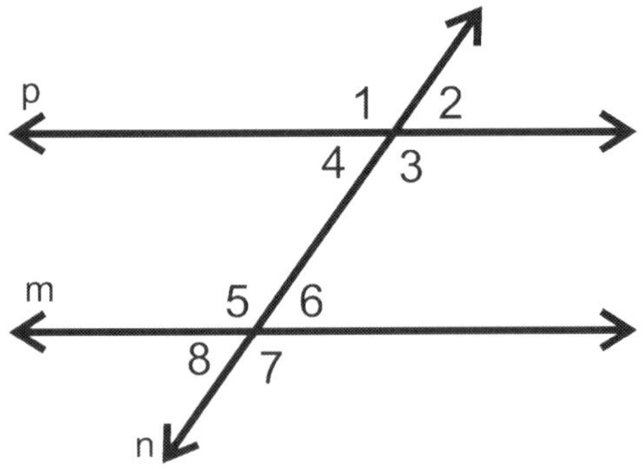

Fig 9

In Fig 8, $\overline{AB}$ and $\overline{EF}$ **are not parallel**, the corresponding angles ∠a and ∠e ; ∠c and ∠g ; ∠b and ∠f ; ∠d and ∠h are **not equal**.

In Fig 9, Lines p and m **are parallel**, the corresponding angles ∠1 and ∠5 ; ∠4 and ∠8 ; ∠2 and ∠6 ; ∠3 and ∠7 **are equal**.

6. Alternate Angles

Two angles, not adjoining one another, that are formed on opposite sides of a line that intersects two other lines. If the original two lines are parallel, the alternate angles are equal.
one of a pair of angles with different vertices and on opposite sides of a transversal at its intersection with two other lines:

1. Alternate Interior Angles are a pair of angles on the inner side (inside) of each of those two intersected lines but on opposite sides of the transversal. If two parallel lines are cut by a transversal, the alternate interior angles are congruent Examples of Alternate Interior Angles In the figure shown, l is the transversal that cut the pair of lines. Angles 3 and 4 and angles 1 and 2 are alternate interior angles.

2: **Alternate Exterior Angles** are a pair of angles on the outer side (outside) of each of those two intersected lines but on opposite sides of the transversal.

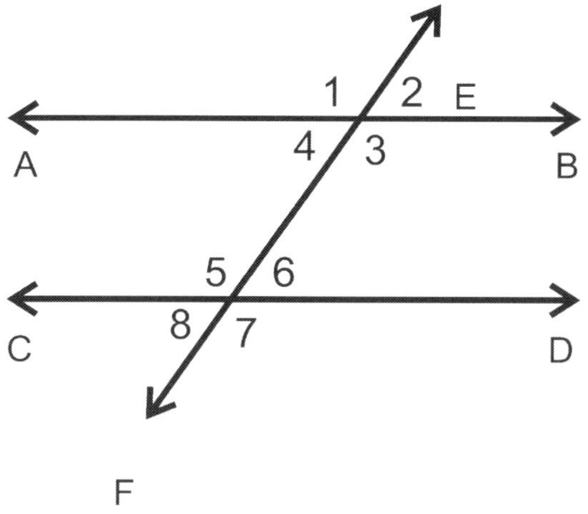

Fig 10

$\angle 3, \angle 4, \angle 5, \angle 6$ Interior Angles

$\angle 1, \angle 2, \angle 7, \angle 8$ Exterior Angles

$\angle 4, \angle 5$; $\angle 3, \angle 6$ } Alternate Interior Angles

$\angle 1, \angle 8$; $\angle 2, \angle 7$ } Alternate Exterior Angles

$\angle 1, \angle 5$; $\angle 2, \angle 6$; $\angle 3, \angle 7$; $\angle 4, \angle 8$ } Corresponding Angles

7. Triangles:

Sum of the angles in any triangle are equal to 180 degrees.

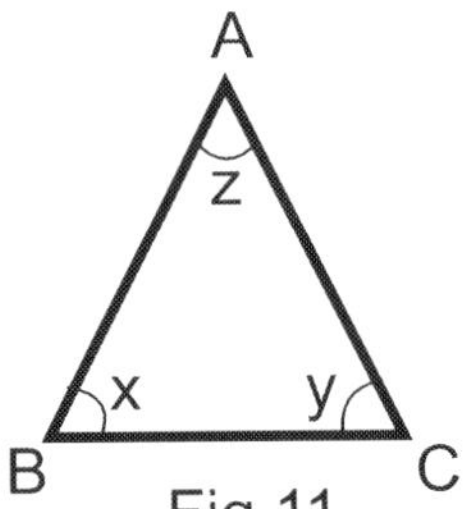

Fig 11

$\angle x + \angle y + \angle z = 180^0$

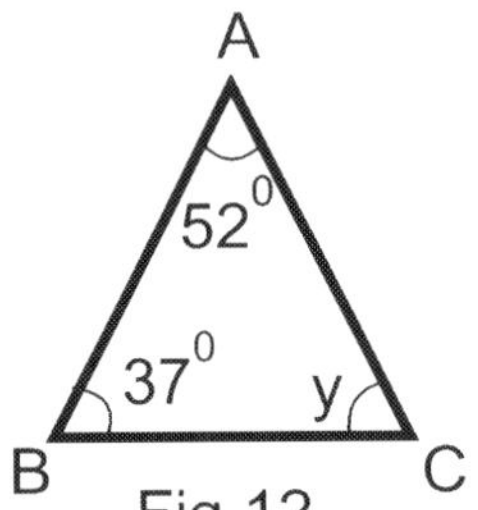

Fig 12

$52^0 + 37^0 + \angle y = 180^0$

$89^0 + \angle y = 180^0$

$\angle y = 180^0 - 89^0$

$\angle y = 91^0$

8. Quadrilaterals: Sum of the angles in any quadrilateral are equal to 360 degrees.

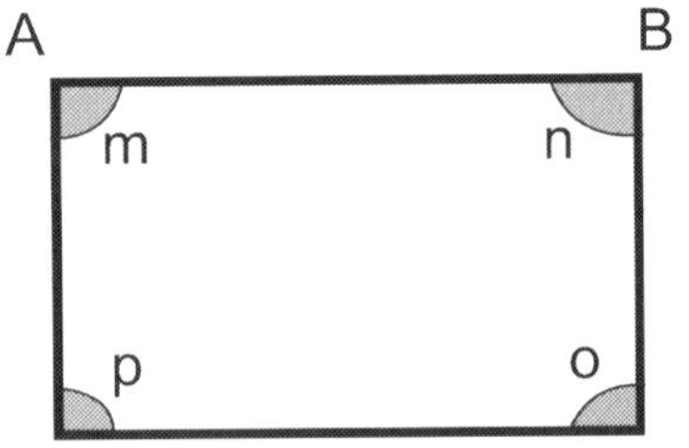

Fig 13

$\angle m + \angle n + \angle o + \angle p = 180^0$

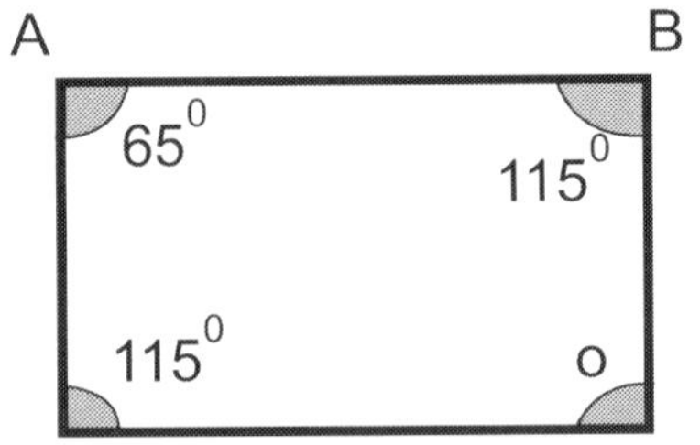

Fig 14

$\angle m + \angle n + \angle o + \angle p = 180^0$

$115^0 + 115^0 + 65^0 + \angle p = 180^0$

$295^0 + \angle p = 180^0$

$\angle p = 180^0 - 295^0$

$\angle p = 65^0$

9. Area of a triangle :

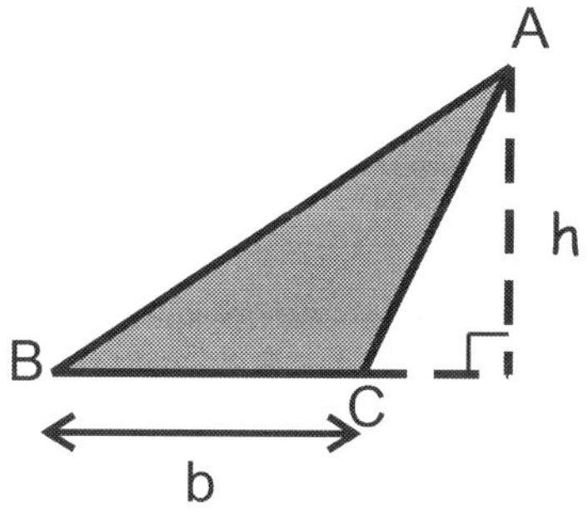

Fig 15

$$A = \frac{1}{2} b h$$

A = Area of the triangle
b = length of the base
h = height

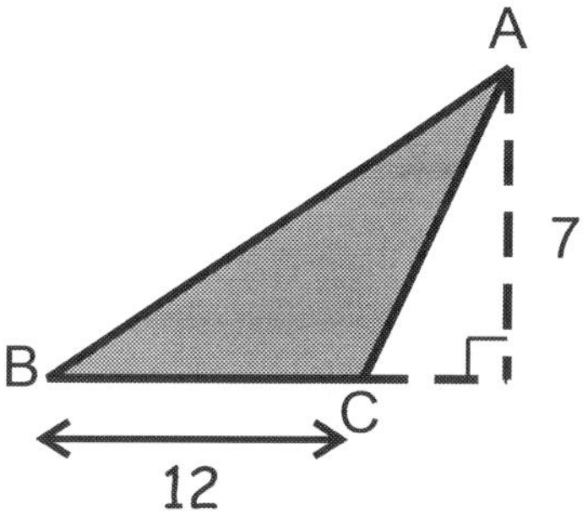

Fig 16

$$A = \frac{1}{2} b h$$

$$A = \frac{1}{2} \times (12)(7)$$

$$= (6)(7)$$

$$= 42 \text{ sq. units}$$

10. Perimeter of a triangle :

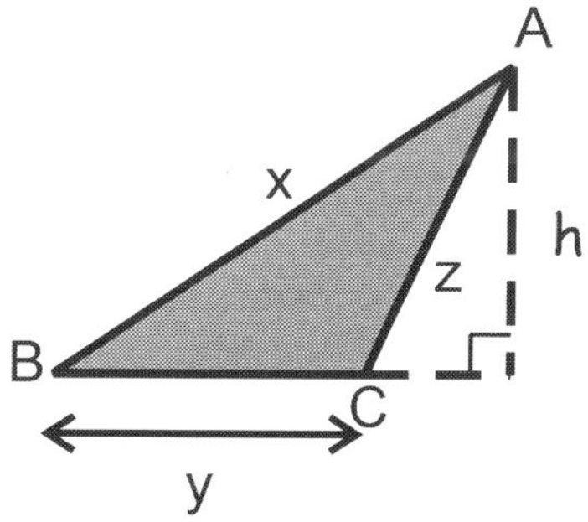

Fig 17

P = x + y + z
P = Perimeter
x, y, z are lengths of the sides of the triangle

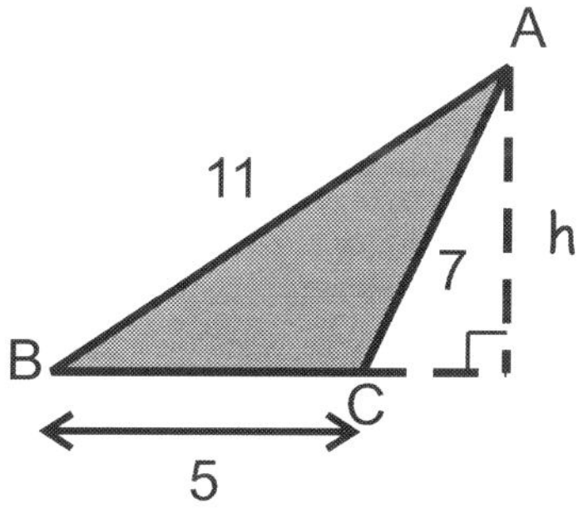

Fig 18

P = x + y + z
P = 11 + 5 + 7
P = 23 units

11. Perimeter and Area of a Square

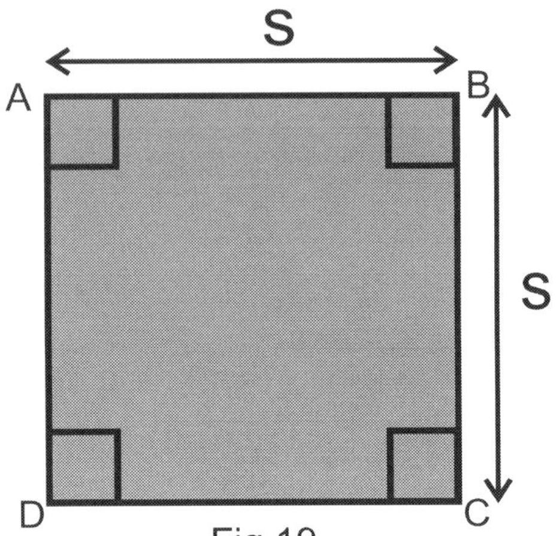

Fig 19

$P = 4s$

$A = s^2$

P = Perimeter
A = Area
s = Length of the side of the square

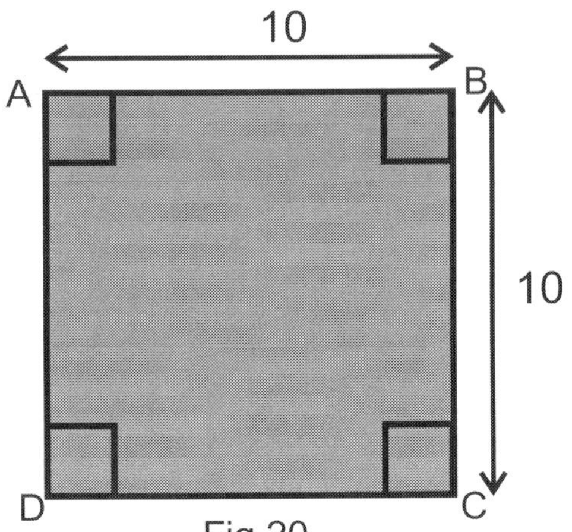

Fig 20

$P = 4s$

$A = s^2$

$P = 4s$

$P = 4 \times 10$

P = 40 units

$A = s^2$

$A = 10^2$

A = 100 sq.units

12. Perimeter and Area of a Rectangle

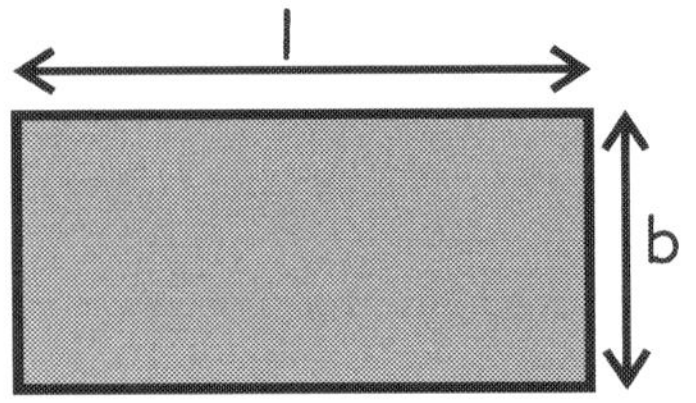

Fig 21

$A = l \times b$

$P = 2(l + b)$

A = Area
P = Perimeter
l = length of the rectangle
b = width of the rectangle

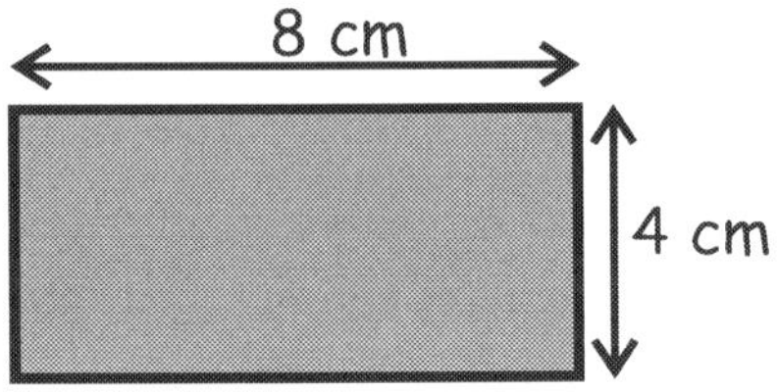

Fig 22

$A = 8 \times 4$

A = 32 sq.cm

$P = 2(8 + 4)$

$P = 2(12)$

P = 24 cm

13. Area of a Trapezium

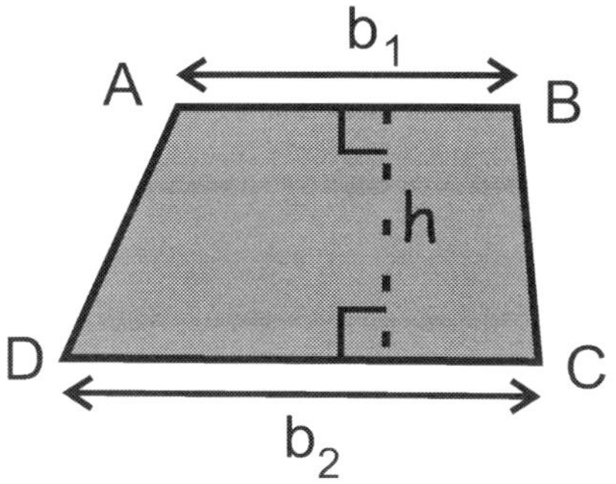

Fig 23

$A = \frac{1}{2} h(b_1 + b_2)$

A = Area
b_1 , b_2 are the lengths of parallel sides
h = Distance between the parallel sides

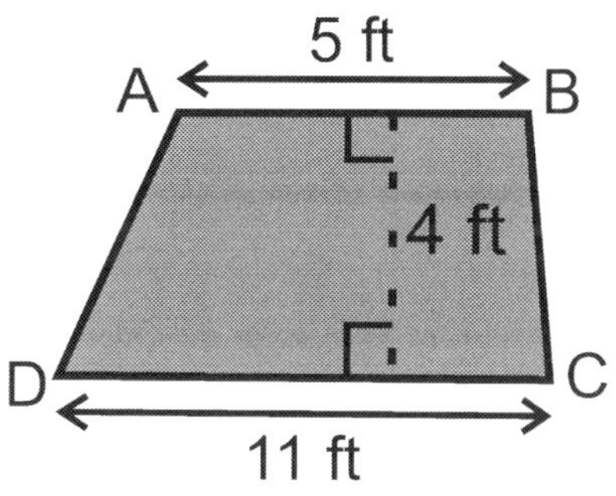

Fig 24

$A = \frac{1}{2} h(b_1 + b_2)$

$A = \frac{1}{2} 4(11 + 5)$

$A = 2(16)$

A = 32 sq.ft

14. Perimeter of a Trapezium

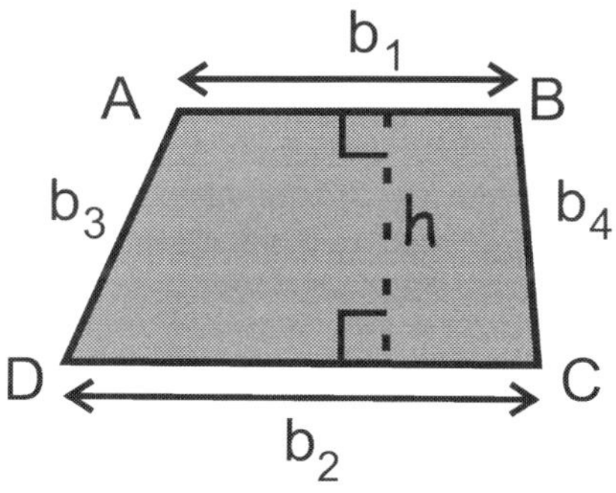

Fig 25

$P = b_1 + b_2 + b_3 + b_4$

P = Perimeter
b_1 , b_2 are the lengths of parallel sides
b_3 , b_4 are the lengths of non parallel sides

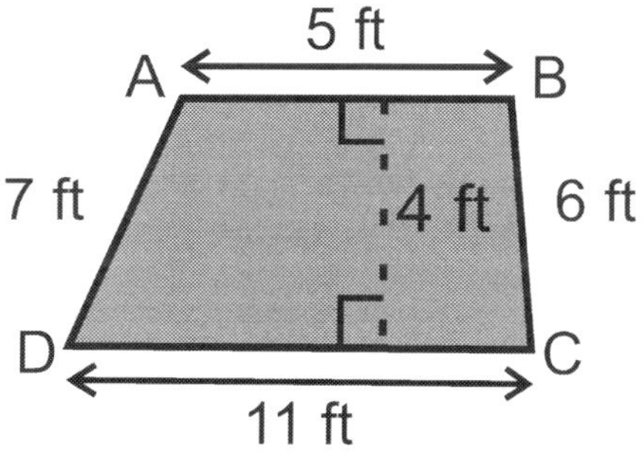

Fig 26

$P = b_1 + b_2 + b_3 + b_4$

$P = 5 + 11 + 7 + 6$

P = 29 ft

15. Area of a parellelogram

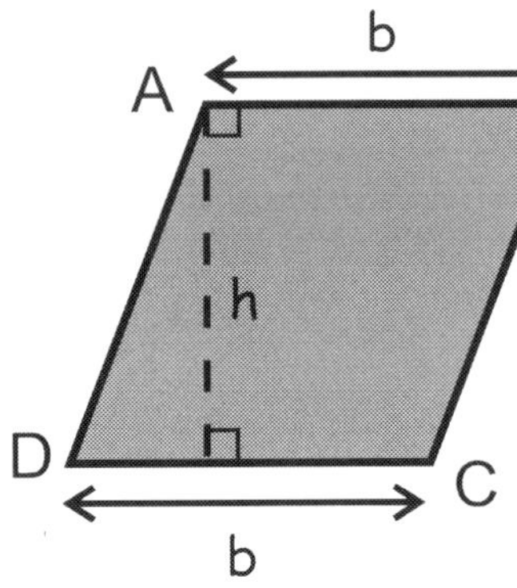

Fig 27

A = bh

A = Area

b = base

h = Height

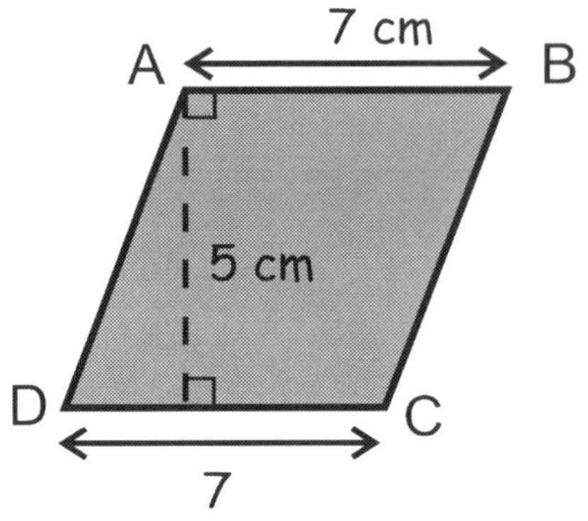

Fig 28

A = bh

A = 5 X 7
A = 45 sq. cm

16. Perimeter of a parellelogram

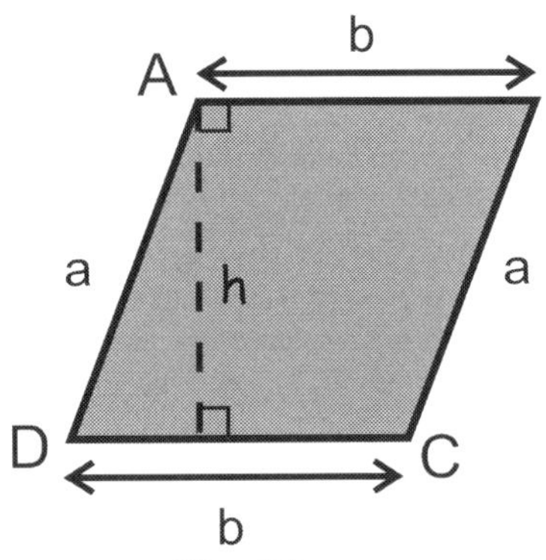

Fig 29

P = a + a + b + b
= 2(a + b)

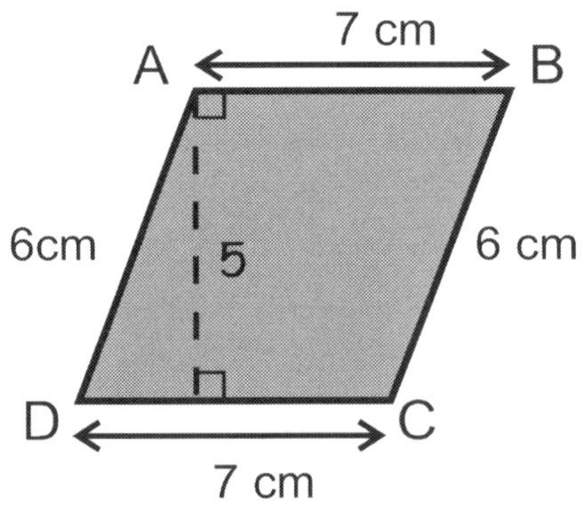

Fig 30

P = 6 + 6 + 7 + 7
= 2(6 + 7)
= 2(13)
= 26 cm

17. Circumference and Area of a Circle

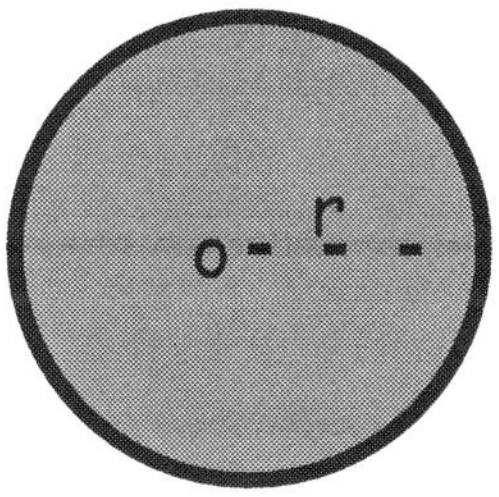

Fig 31

$C = 2\Pi r$

$A = \Pi r^2$

pi

$\Pi = 3.14$

$\Pi = \frac{22}{7}$

C = Circumference of the circle

A = Area of the circle
r = radius

Note : Diameter(d) = 2r

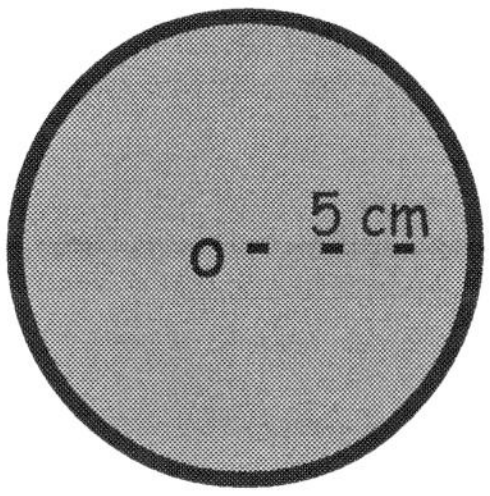

Fig 32

$C = 2\Pi r$

C = 2 X 3.14 X 5

C = 10 X 3.14

C = 31.4 cm

A = 3.14 X 5 X 5
A = 3.14 X 25
A = 78.5 sq.cm

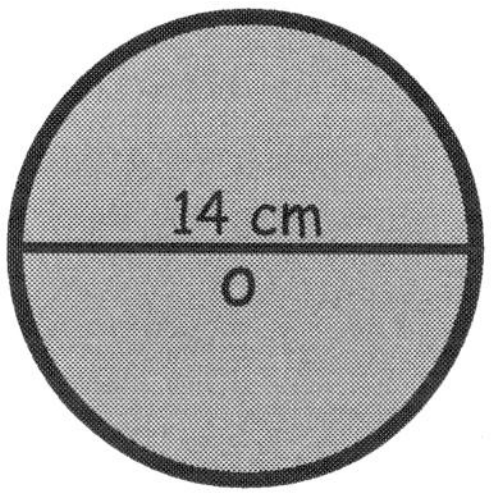

Fig 33

Diameter(d) = 14 cm

Radius $= \frac{14}{2}$

Radius = 7 cm

C = 2 X 3.14 X 7

C = 14 X 3.14

C = 43.86cm

A = 3.14 X 7 X 7
A = 3.14 X 49
A = 153.86 sq.cm

18. Volume of a sphere

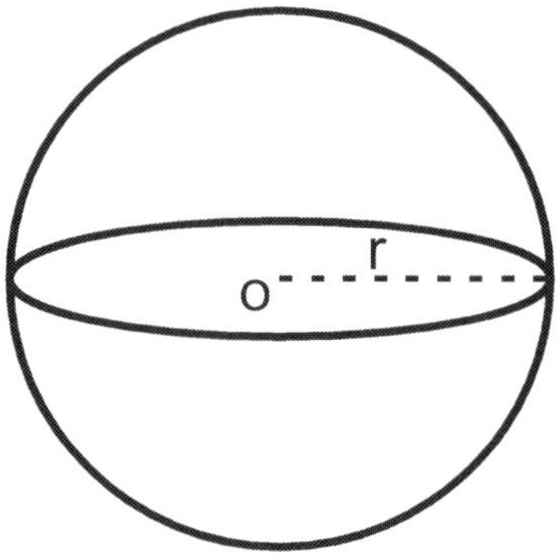

Fig 34

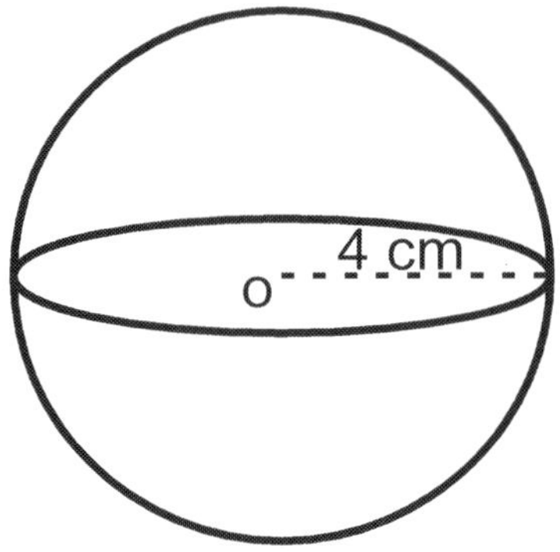

Fig 35

$V = \frac{4}{3} \pi r^3$

V = Volume of the sphere
r = Radius of the sphere

Diameter = Twice the radius

pi

$\pi = 3.14$

$\pi = \frac{22}{7}$

$V = \frac{4}{3} \pi r^3$

$V = \frac{4}{3} \times 3.14 \times 4^3$

$V = 268.08 \text{ cm}^3$

19. Surface area of a sphere

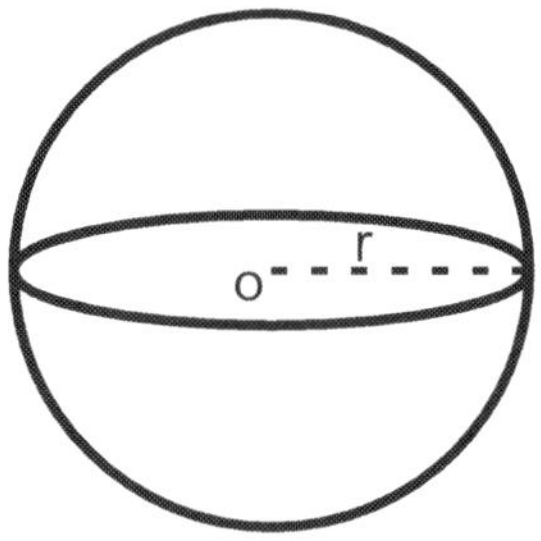

Fig 36

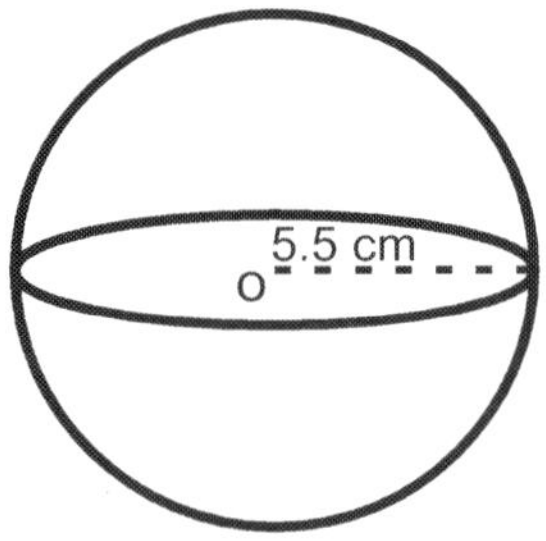

Fig 37

$S = 4 \pi r^2$

pi

$\pi = 3.14$

$\pi = \frac{22}{7}$

$S = 4 \pi r^2$

$S = 4 \times 3.14 \times (5.5)^2$

$S = 379.94 \text{ cm}^2$

20. Volume of a cone

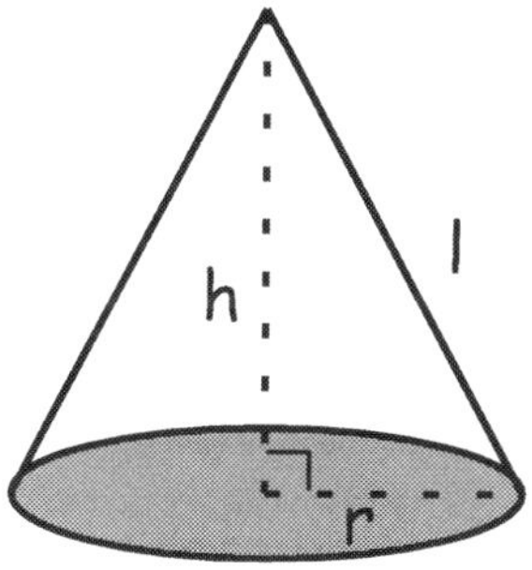

Fig 38

$$V = \frac{1}{3} \Pi r^2 h$$

V = Volume of the cone
r = Radius
h = Height
l = Slant height

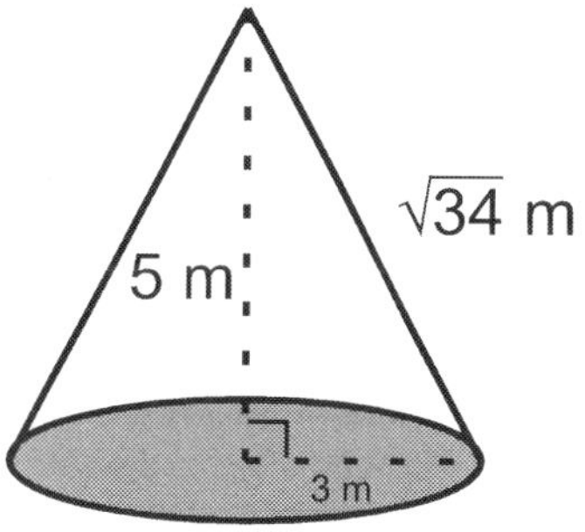

Fig 39

$$V = \frac{1}{3} \Pi r^2 h$$

$$V = \frac{1}{3} X 3.14 X 3^2 X 5$$

$$V = 47.1239 \, m^3$$

21. Lateral surface area of a cone

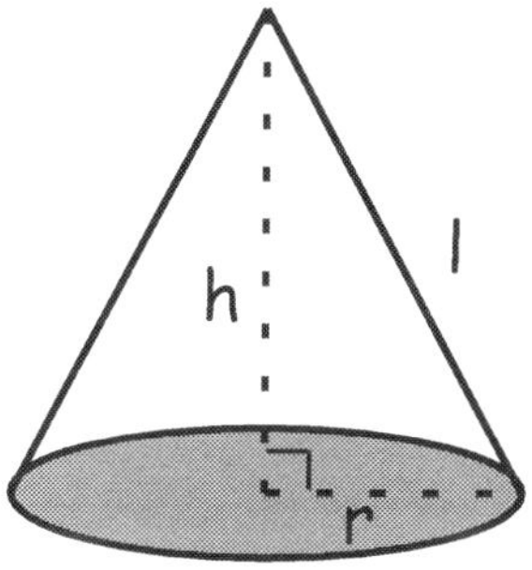

Fig 40

L.S.A $= \Pi rl$

$= \Pi r X \sqrt{r^2 + h^2}$

$l = \sqrt{r^2 + h^2}$

L.S.A = Lateral surface area
l = Slant height of the cone
r = radius
h = height

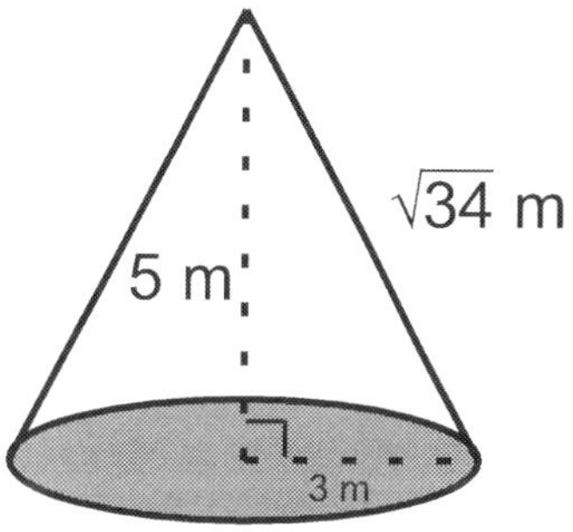

Fig 41

L.S.A $= \Pi rl$

L.S.A $= 3.14 X 3 X \sqrt{34}$

$= 54.9554 \, m^2$

22. Base surface area of a cone

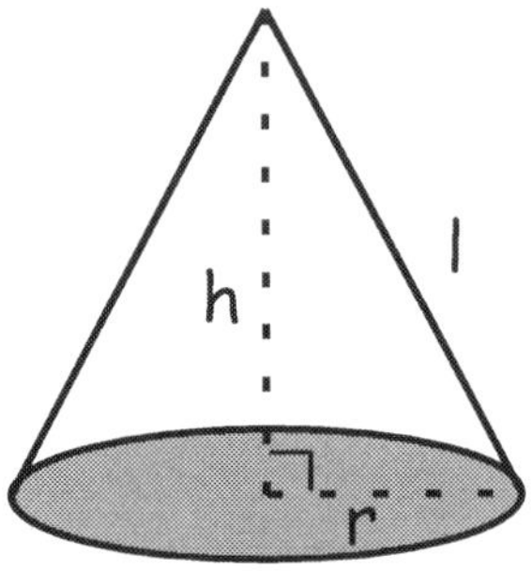

Fig 42

$B.S.A = \prod r^2$

B.S.A = Base surface area
r = radius
h = height

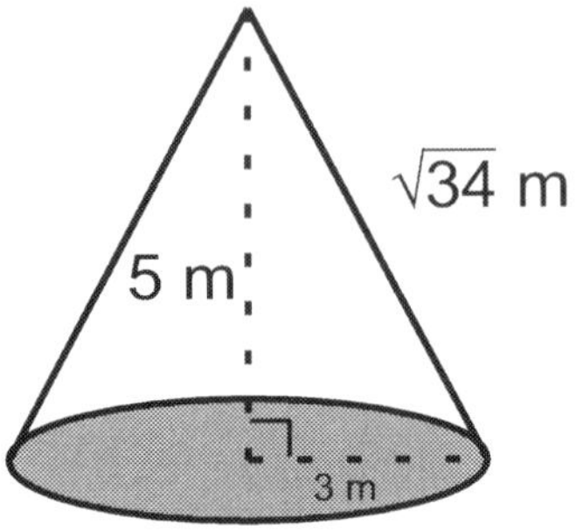

Fig 43

$B.S.A = \prod r^2$

$B.S.A = 3.14 \times 3^2$

$B.S.A = 28.2743\ m^2$

23. Total surface area of a cone

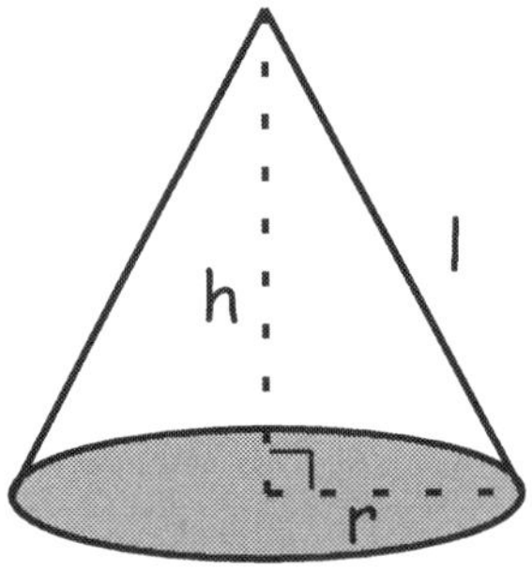

Fig 44

T.S.A = L.S.A + B.S.A

$T.S.A = \prod rl + \prod r^2$

$T.S.A = \prod r(l + r)$

$T.S.A = \prod r(r + \sqrt{r^2 + h^2})$

T.S.A = Total surface area
r = radius
h = height

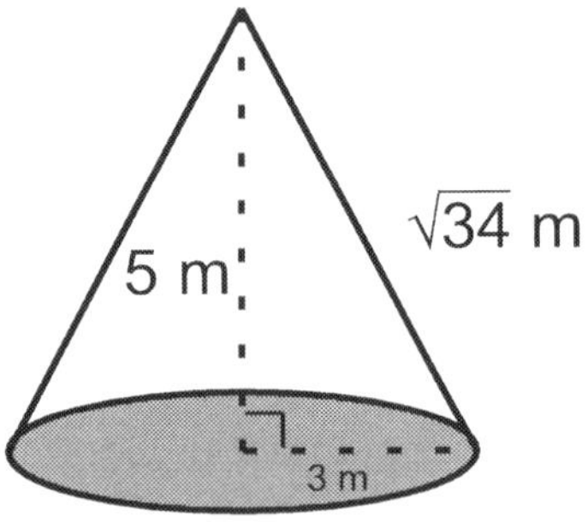

Fig 45

$T.S.A = \prod r(l + r)$

$T.S.A = 3.14 \times 3\ (\sqrt{34} + 3)$

$T.S.A = 83.2298\ m^2$

24. Volume of a Cylinder

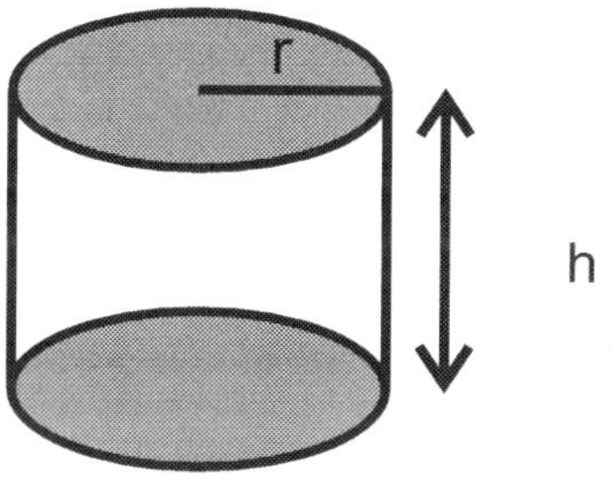

Fig 46

$V = \pi r^2 h$

V = Volume
r = radius
h = height

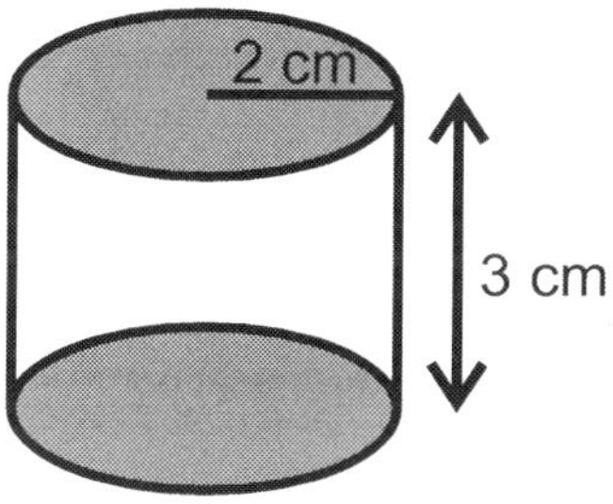

Fig 47

$V = \pi r^2 h$

$V = 3.14 \times 2^2 \times 3$

$V = 37.6991 \text{ cm}^3$

25. Lateral Surface area of a Cylinder

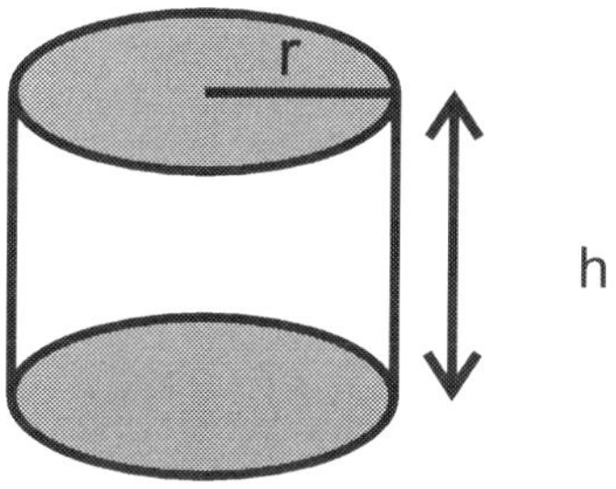

Fig 48

$L.S.A = 2 \pi r h$

L.S.A = Lateral surface area
r = radius
h = height

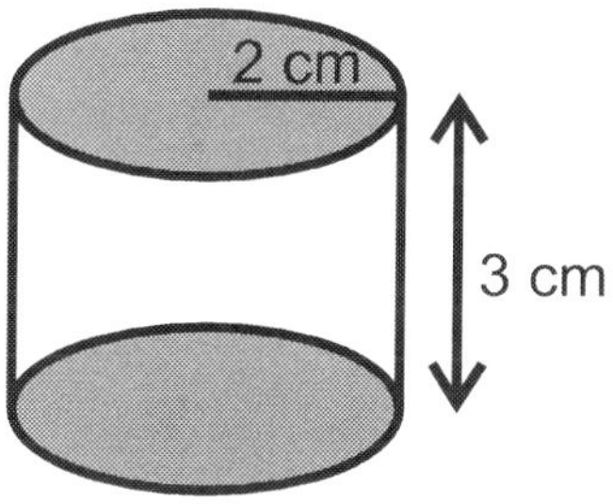

Fig 49

$L.S.A = 2 \pi r h$

$L.S.A = 2 \times 3.14 \times 2 \times 3$

$L.S.A = 37.6991 \text{ cm}^2$

26. Top and bottom Surface area of a Cylinder

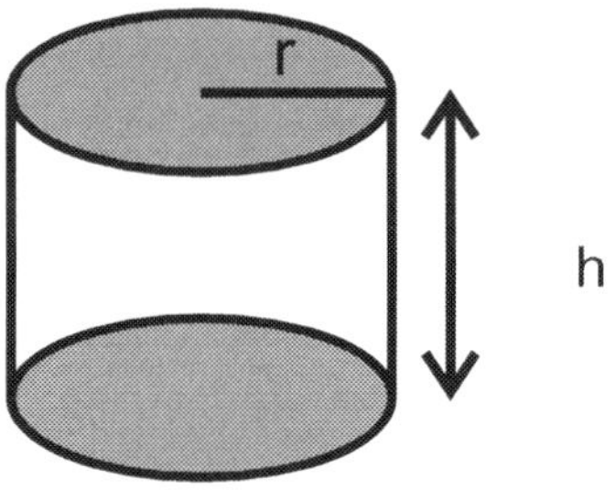

Fig 50

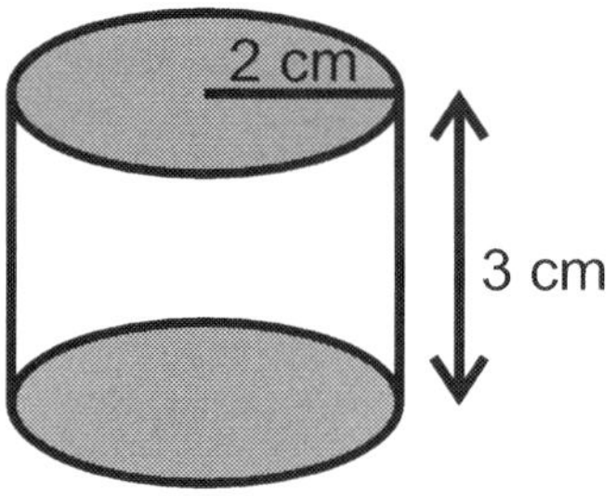

Fig 51

B.S.A $= \pi r^2$

B.S.A = Bottom surface area
r = radius
h = height

B.S.A $= \pi r^2$

B.S.A $= 3.14 \times 2^2$

B.S.A $= 12.5664\ cm^2$

> Top and bottom surface area of the cylinder are equal. While calculating total surface area of the cylinder, remember to add the bottom surface are of the cylinder twice

T.S.A = Total surface area
T.S.A = L.S.A + B.S.A + B.S.A

T.S.A $= 2\pi rh + \pi r^2 + \pi r^2$

T.S.A $= 2\pi rh + 2\pi r^2$

T.S.A $= 2\pi r\,(h + r)$

T.S.A $= 2\pi r\,(h + r)$

T.S.A $= 2 \times 3.14 \times 2\ (3 + 2)$

T.S.A $= 62.8318\ cm^2$

27. Volume of a Cuboid

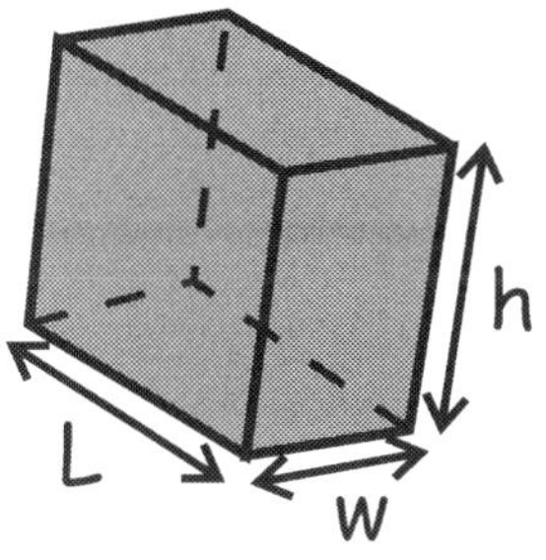

Fig 52

V = lwh
V = Volume
l = Length
w = Width or breadth
h = Height

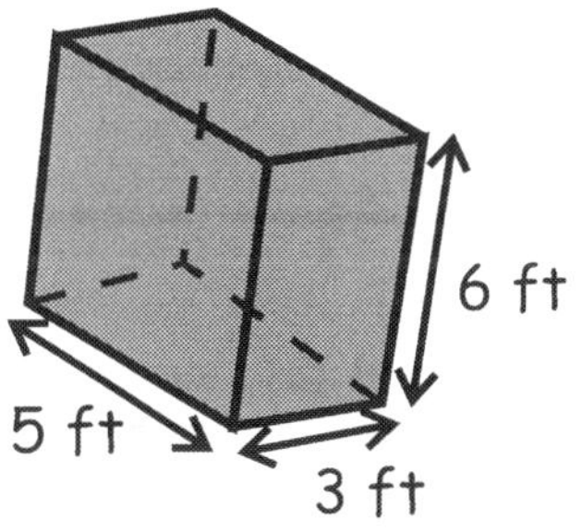

Fig 53

V = lwh

V = 5 X 3 X 6

V = 90 ft^3

28. Surface area of a Cuboid

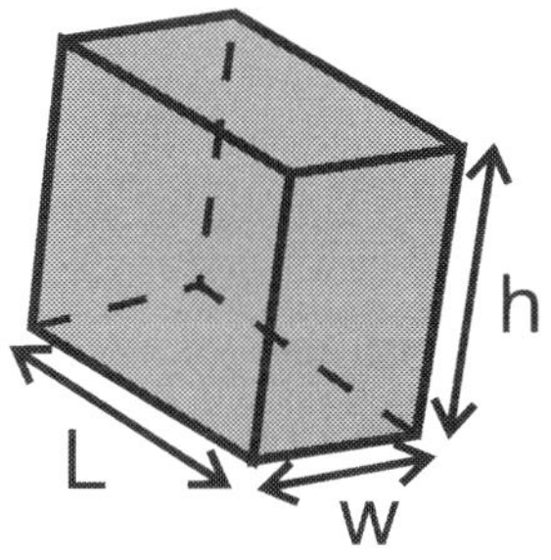

Fig 54

T.S.A = 2(lw + lh + wh)
T.S.A = Total surface area
l = Length
w = Width or breadth
h = Height

L.S.A = 2(lh + wh)
L.S.A = Lateral surface area
l = Length
w = Width or breadth
h = Height

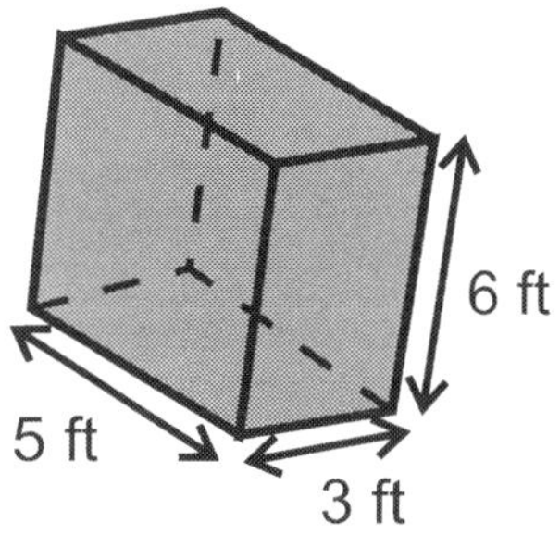

Fig 55

T.S.A = 2(lw + lh + wh)

T.S.A = 2(5X3 + 5X6 + 3X6)

T.S.A = 126 ft^2

L.S.A = 2(lh + wh)

L.S.A = 2(5X6 + 3X6)

L.S.A = 96 ft^2

29. Volume of a Cube

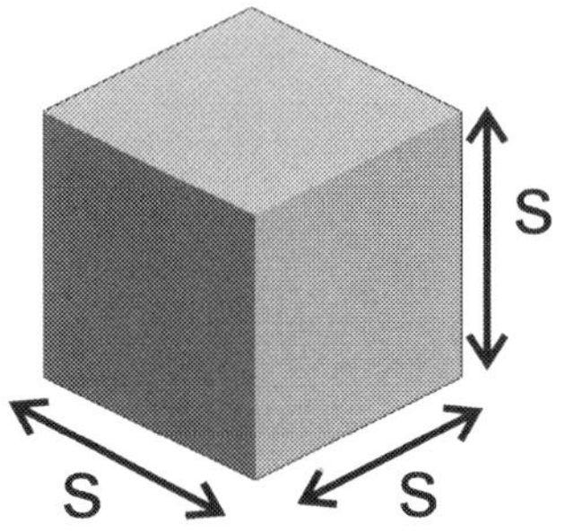

Fig 56

$V = s^3$

V = Volume

s = Side length of the cube

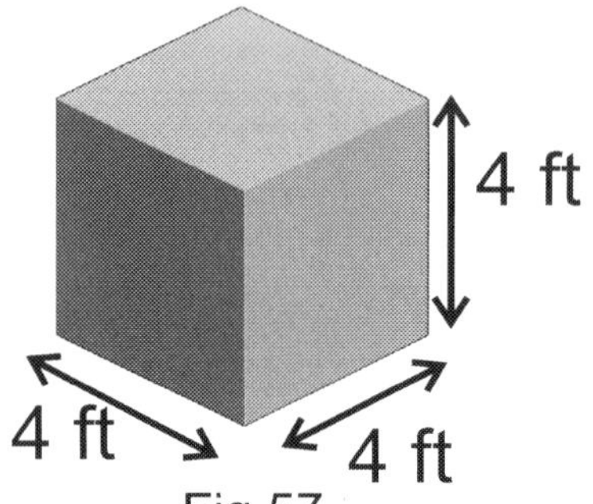

Fig 57

$V = s^3$

$V = 4^3$

$V = 64\ ft^3$

30. Surface area of a Cube

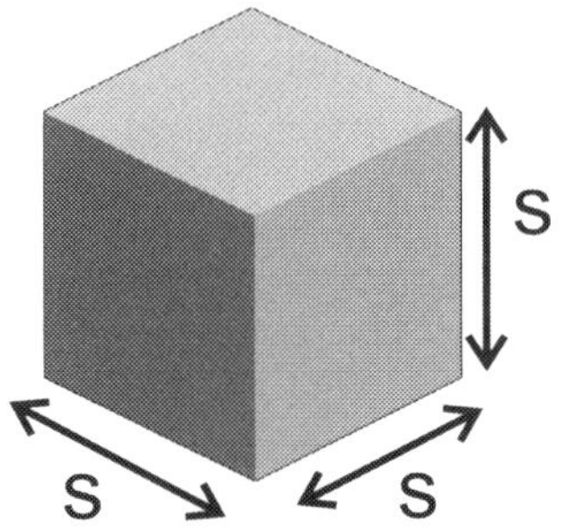

Fig 58

$T.S.A = 6s^2$

T.S.A = Total surface area

s = Side length of the cube

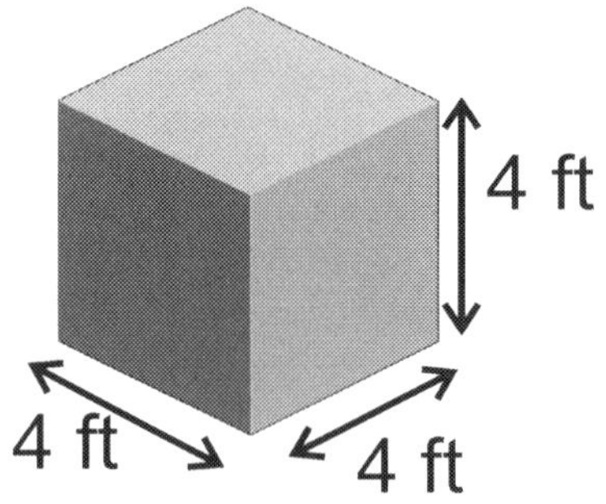

Fig 59

$T.S.A = 6s^2$

$T.S.A = 6 \times 4^2$

$T.S.A = 96\ ft^2$

31. Volume of a square Pyramid

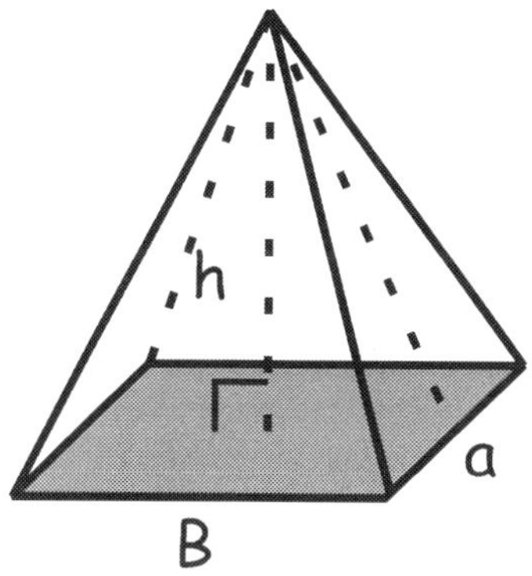

Fig 60

$$V = \frac{1}{3} B h$$

V = Volume
B = Base area
h = Height

$$V = \frac{1}{3} a^2 h$$

V = Volume
a = side length
h = Height

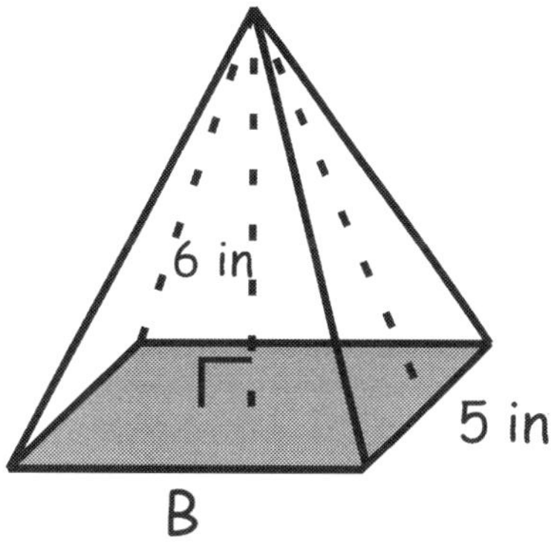

Fig 61

$$V = \frac{1}{3} a^2 \ h$$

$$V = \frac{1}{3} \times 5^2 \times 6$$

$$V = \frac{1}{3} \times 150$$

$$V = 50 \text{ in}^3$$

32. Lateral surface area of a square Pyramid

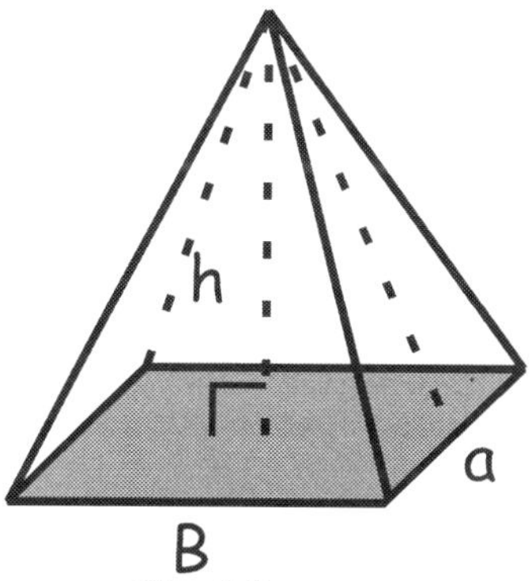

Fig 62

$$S.A = a \times \sqrt{a^2 + 4h^2}$$

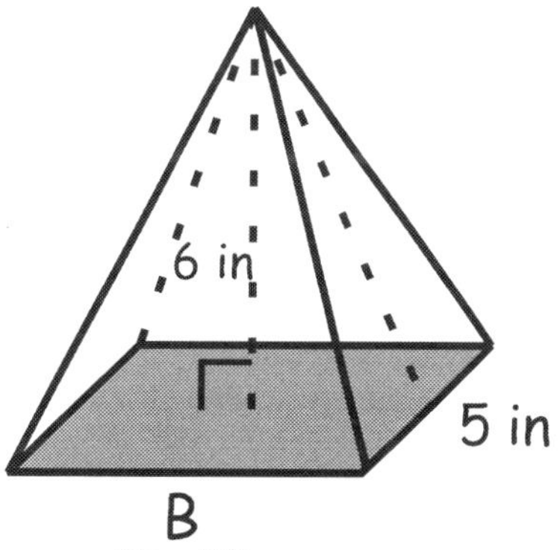

Fig 63

$$S.A = a \times \sqrt{a^2 + 4h^2}$$

$$S.A = 5 \times \sqrt{5^2 + 4 \times 6^2}$$

$$S.A = 65 \text{ in}^2$$

32. Base surface area of a square Pyramid

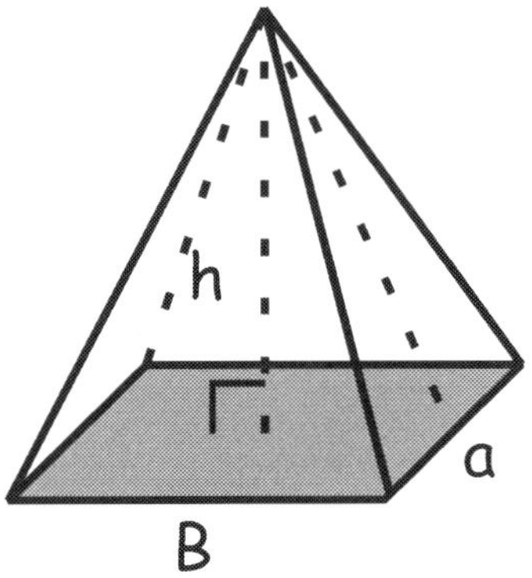

Fig 64

$B.S.A = a^2$

$T.S.A = a^2 + a \times \sqrt{a^2 + 4h^2}$

B.S.A = Base surface area
T.S.A = Total surface area
T.S.A = B.S.A + L.S.A

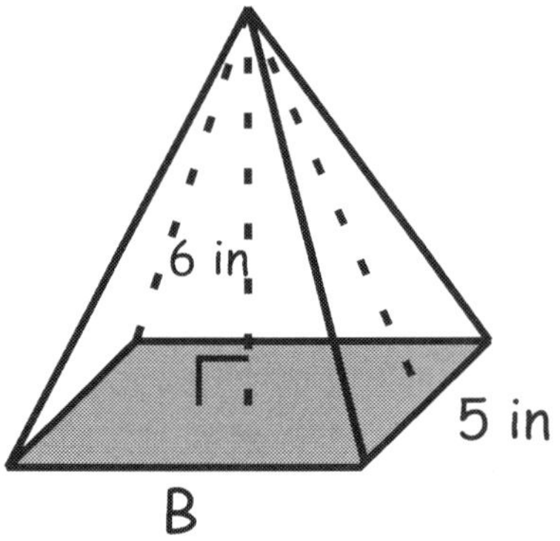

Fig 65

$B.S.A = 5 \times 5$
$B.S.A = 25\ in^2$

$T.S.A = 5^2 + 5 \times \sqrt{5^2 + 4 \times 6^2}$

$T.S.A = 25 + 65$

$T.S.A = 90\ in^2$

32. Volume of a Triangular prism

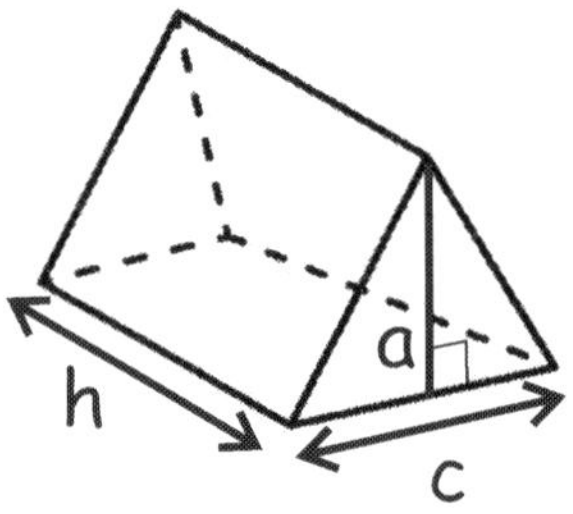

Fig 66

$V = Bh$

$V = \frac{1}{2}ach$

V = Volume
a = apothem
h = height
c = base length of the triangle

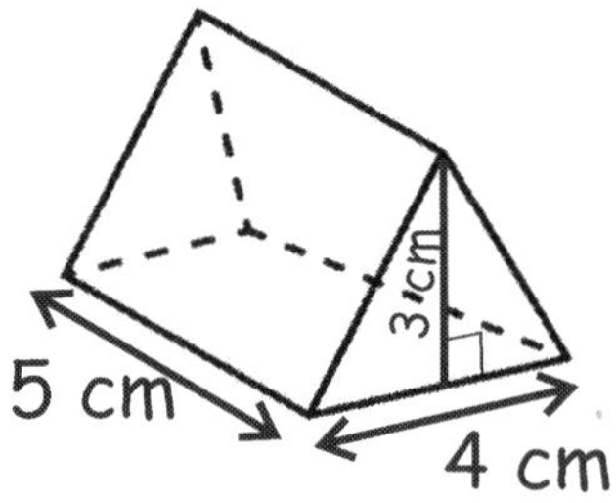

Fig 67

$V = Bh$

$V = \frac{1}{2}ach$

$V = \frac{1}{2}(3 \times 4 \times 5)$

$V = 30\ cm^3$

32. Volume of a triangular pyramid

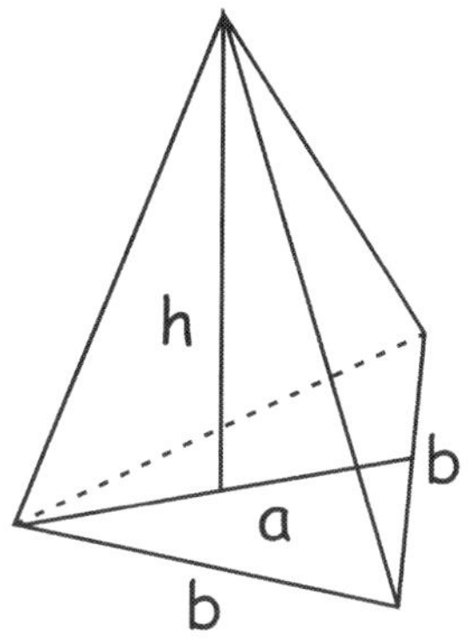

Fig 68

Volume of Triangular pyramid $= \frac{1}{6}$ abh

a = Apothem length of the pyramid
b = base length of the pyramid
h = Height of the pyramid

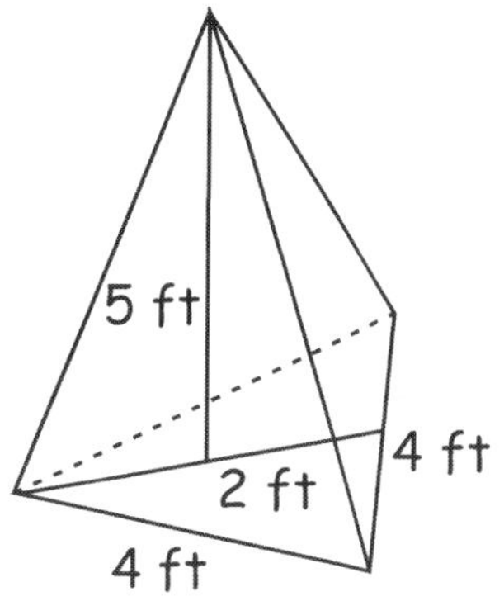

Fig 69

Volume of Triangular pyramid $= \frac{1}{6}$ abh

Volume of Triangular pyramid $= \frac{1}{6}$ (2X4X5)

Volume of Triangular pyramid $= \frac{1}{6}$ (40)

Volume of Triangular pyramid = 6.66 ft^3

33. Volume of a pentagonal pyramid

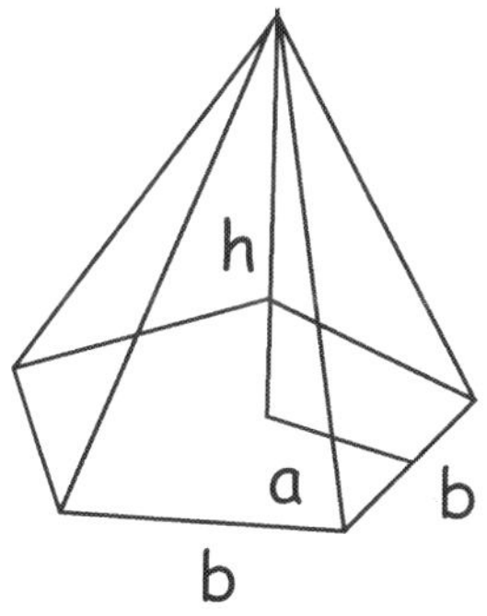

Fig 70

Volume of pentagonal pyramid $= \frac{5}{6}$ abh

a = Apothem length of the pyramid
b = base length of the pyramid
h = Height of the pyramid

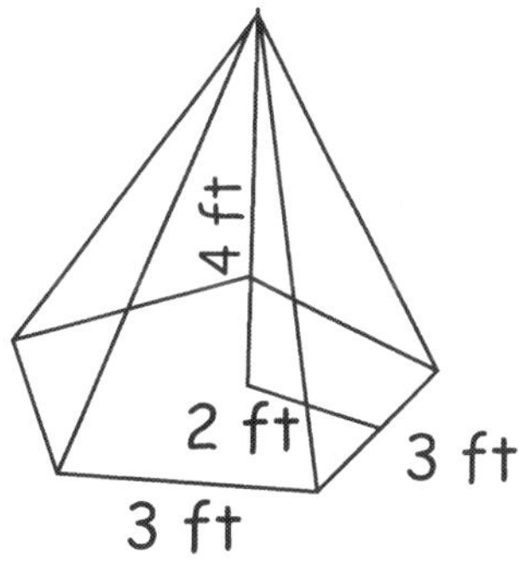

Fig 71

Volume of pentagonal pyramid $= \frac{5}{6}$ abh

Volume of pentagonal pyramid $= \frac{5}{6}$ (2X3X4)

Volume of pentagonal pyramid $= \frac{5}{6}$ (24)

Volume of pentagonal pyramid = 20 ft^3

34. Volume of a hexagonal pyramid

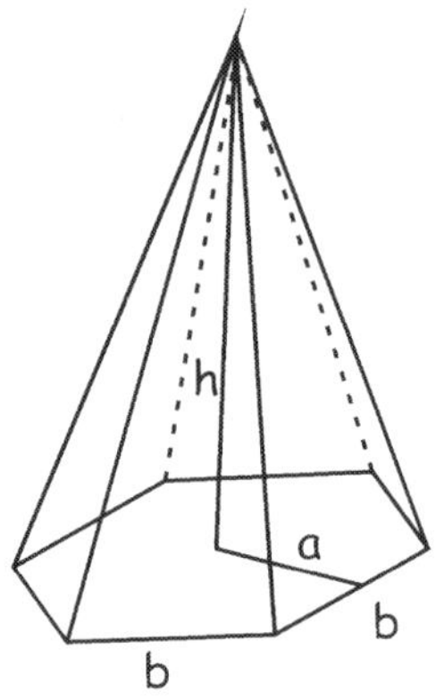

Fig 72

Volume of Hexagonal pyramid = abh

a = Apothem length of the pyramid
b = base length of the pyramid
h = Height of the pyramid

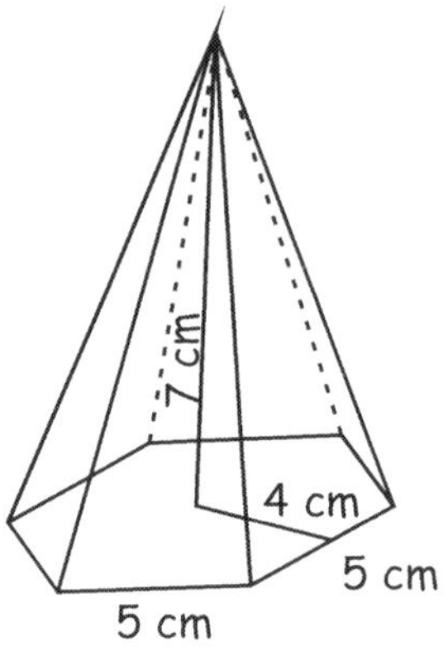

Fig 73

Volume of Hexagonal pyramid = 5X4X7
Volume of Hexagonal pyramid = 140 cm^3

Number series

Sequences of numbers which follow specific patterns are called progression. Depending on the pattern, the progressions are classified as follows.

(i) Arithmetic Progression

(ii) Geometric Progression and

(iii) Harmonic Progression

ARITHMETIC PROGRESSION (A.P.)

Numbers (or terms) are said to be in arithmetic progression when each one, except the first, is obtained by adding a constant to the previous number (or term).

An arithmetic progression can be represented by a, a + d, a + 2d, ..., [a + (n – 1)d]. Here, d is added to any term to get the next term of the progression. The term a is the first term of the progression, n is the number of terms in the progression and d is the common difference. The nth term is normally represented by T_n and the sum to n terms of an A.P. is denoted by S_n

n^{th} term = $T_n = a + (n - 1)d$

Sum to n terms = $S_n = \left(\frac{n}{2}\right)[2a + (n - 1)d]$

The sum to n terms of an A.P. can also be written in a different manner. That is,

sum of n terms = $\left(\frac{n}{2}\right)[2a + (n - 1)d] = \left(\frac{n}{2}\right)[a + \{a + (n - 1)d\}]$

But, when there are n terms in an A.P., a is the first term and $\{a + (n - 1)d\}$ is the last term. Hence,

$S_n = \left(\frac{n}{2}\right)$[first term + last term]

The average of all the terms in an A.P. is called the arithmetic mean (A.M.) of the A.P. Since the average of a certain numbers is equal to the {sum of all the number/number of numbers}.

A.M. of n terms in A.P. $= \frac{S_n}{n} = \frac{1}{n}\left(\frac{n}{2}\right)$

(First Term + Last Term) $= \frac{\text{(First Term + Last Term)}}{2}$

i.e. The A.M. of an A.P. is the average of the first and the last terms of the A.P.

The A.M. of an A.P. can also be obtained by considering any two terms which are EQUIDISTANT from the two ends of the A.P. and taking their average, i.e.

(a) the average of the second term from the beginning and the second term from the end is equal to the A.M. of the A.P.

(b) the average of the third term from the beginning and the third term from the end is also equal to the A.M. of the A.P. and so on.

In general, the average of the kth term from the beginning and the kth term from the end is equal to the A.M. of the A.P.

If the A.M. of an A.P. is known, the sum to n terms of the series (S_n) can be expressed as $S_n = n$ (A.M.)

In particular, if three numbers are in arithmetic progression, then the middle number is the A.M. i.e. if a, b and c are in A.P., then b is the A.M. of the three terms and

$b = \frac{a + c}{2}$.

If a and b are any two numbers, then their A.M. $= \frac{a + b}{2}$.

Note:

(i) If three numbers are in A.P., we can take the three terms to be (a – d), a and (a + d).

(ii) If four numbers are in A.P., we can take the four terms to be (a – 3d), (a – d), (a + d) and (a + 3d). The common difference in this case is 2d and not d.

(iii) If five numbers are in A.P., we can take the five terms to be (a – 2d), (a – d), a, (a + d) and (a + 2d).

Inserting arithmetic mean between two numbers:

When n arithmetic means $a_1, a_2, \ldots\ldots\ldots\ldots, a_n$ are inserted between a and b, then $a, a_1, a_2, \ldots\ldots\ldots, a_n, b$ are in A.P.

$t_1 = a$ and $t_{n+2} = b$ of A.P.

The common difference of the A.P. can be obtained as follows:

Given that, n arithmetic means are there between a and b.

$\therefore a = t_1$ and $b = t_{n+2}$

Let d be the common difference.

$b = t_1 + (n+1)\,d$

$b = a + (n+1)d$

$$d = \frac{(b-a)}{(n+1)}$$

Some important results:

The sum to n terms of the following series are quite useful and, hence, should be remembered by students.

(i) Sum of the first n natural numbers $= \sum_1^n i = \frac{n(n+1)}{2}$

(ii) Sum of squares of the first n natural numbers $= \sum_1^n i^2 = \frac{n(n+1)(2n+1)}{6}$

(iii) Sum of cubes of first n natural numbers

Example 1:

Find the 14th term of an A.P. whose first term is 3 and the common difference is 2.

Solution:

The nth term of an A.P. is given by $t_n = a + (n - 1)d$, where a is the first term and d is the common difference.

$\therefore t_{14} = 3 + (14 - 1)\,2 = 29$

Example 2:

Find the first term and the common difference of an A.P. if the 3rd term is 6 and the 17th term is 34.

Solution:

If a is the first term and the common difference d, then we have

$a + 2d = 6$ ------- (1)

$a + 16d = 34$ ----- (2)

On subtracting equation (1) from equation (2), we get

$14d = 28 \quad d = 2$

Substituting the value of d in equation (1), we get $a = 2$

$\therefore a = 2$ and $d = 2$

Example 3:

Find the sum of the first 22 terms of an A.P. whose first term is 4 and the common difference is 4/3.

Solution:

Given that, $a = 4$ and $d = \frac{4}{3}$.

We have $S_n = \frac{n}{2}[2a + (n - 1)d]$

$$S_{22} = \left(\frac{22}{2}\right)\left[(2)(4) + (22 - 1)\left(\frac{4}{3}\right)\right] = (11)(8 + 28) = 396$$

Example 4:

Divide 124 into four parts in such a away that they are in A.P. and the product of the first and the 4th part is 128 less than the product of the 2nd and the 3rd parts.

Solution:

Let the four parts be (a – 3d), (a – d), (a + d) and (a + 3d). The sum of these four parts is 124,

i.e. $4a = 124$ $a = 31$

$(a - 3d)(a + 3d) = (a - d)(a + d) - 128$

$a^2 - 9d^2 = a^2 - d^2 - 128$

$8d^2 = 128$ d = ®4

As a = 31, taking d = 4, the four parts are 19, 27, 35 and 43.

Note:

If d is taken as –4, then the same four numbers are obtained, but in decreasing order.

Example 5:

Find the three terms in A.P., whose sum is 36 and product is 960.

Solution:

Let the three terms of an A.P. be (a – d), a and (a + d).

Sum of these terms is 3a.

$3a = 36$ $a = 12$

Product of these three terms is

$(a + d)\, a\, (a - d) = 960$ $(12 + d)(12 - d) = 80$

$144 - d^2 = 80$ d = ®8

Taking d = 8, we get the terms as 4, 12 and 20.

Note:

If d is taken as –8, then the same numbers are obtained, but in decreasing order.

Geometric Progression (G.P.):

Numbers are said to be in geometric progression when the ratio of any quantity to the number that follows it is the same. In other words, any term of a G.P. (except the first one) can be obtained by multiplying the previous term by a constant factor.

The constant factor is called the common ratio and is normally represented by r. The first term of a G.P. is generally denoted by a.

A geometric progression can be represented by a, ar, ar^2, ... where a is the first term and r is the common ratio of the G.P. n^{th} term of the G.P. is ar^{n-1} i.e. $t_n = ar^{n-1}$

Sum to n terms = $S_n = \frac{a(1 - r^n)}{1 - r} = \frac{a(r^n - 1)}{r - 1} = \frac{r(ar^{n-1}) - a}{r - 1}$

The sum to n terms of a geometric progression can also be written as

$S_n = \frac{r(\text{Last term}) - \text{First term}}{r - 1}$

Note:

If n terms $a_1, a_2, a_3, \ldots\ldots\ldots a_n$ are in G.P., then the geometric mean (G.M.) of these n terms is given

by = $\sqrt[n]{a_1 a_2 a_3 \ldots\ldots a_n}$

If three terms are in geometric progression, then the middle term is the geometric mean of the G.P., i.e. if a, b and c are in G.P., then b is the geometric mean of the three terms.

If there are two terms a and b, their geometric mean is given by G.M. = $\sqrt{ab}$. We see that the 3 terms a, $\sqrt{ab}$, b are in G.P.

For any two positive numbers a and b, their arithmetic mean is always greater than or equal to their

geometric mean, i.e. for any two positive numbers a and b, $\frac{a + b}{2} \geq \sqrt{ab}$.

The equality holds if and only if a = b.

Infinite geometric progression:

If $-1 < r < 1$ (or $|r| < 1$), then the sum of a geometric progression does not increase infinitely but "converges" to a particular value, no matter how many terms of the G.P. we take. The sum of an infinite geometric progression is represented by $S_{\prod}$ and is given by the formula,

$S_{\prod} = \dfrac{a}{1-r^2}$, if $|r| < 1$.

Example 6:

Find the 7th term of the G.P. whose first term is 6 and common ratio is 2/3.

Solution:

Given that, $t_1 = 6$ and $r = \dfrac{2}{3}$

We have $t_n = a \,.\, r^{n-1}$

$$t_7 = (6)\left(\frac{2}{3}\right)^6 = \frac{(6)(64)}{729} = \frac{128}{243}$$

Example 7:

Find the common ratio of the G.P. whose first and last terms are 25 and 1/625 respectively and the sum of the G.P. is 19531/625.

Solution:

We know that the sum of a G.P is $\dfrac{\text{first term} - r(\text{last term})}{1 - r}$

$$\frac{19531}{625} = \frac{25 - (r/625)}{1 - r} \qquad r = 1/5$$

Example 8:

Find three numbers of a G.P. whose sum is 26 and product is 216.

Solution:

Let the three numbers be a/r, a and ar.
Given that,

$a/r \, . \, a \, . \, ar = 216;$

$\Rightarrow a^3 = 216; a = 6$
$a/r + a + ar = 26$
$\Rightarrow 6 + 6r + 6r^2 = 26r$
$\Rightarrow 6r^2 \quad 20r + 6 = 0$
$\Rightarrow 6r^2 \quad 18r \quad 2r + 6 = 0$
$\Rightarrow 6r\,(r \quad 3) \quad 2(r \quad 3) = 0$
$\Rightarrow r = 1/3$ (or) $r = 3$

Hence the three numbers are 2, 6 and 18 (or) 18, 6 and 2

Example 9:

If $|\,x\,| < 1$, then find the sum of the series $2 + 4x + 6x^2 + 8x^3 + \ldots$

Solution:

Let $S = 2 + 4x + 6x^2 + 8x^3 +$ ------- (1)

$xS = 2x + 4x^2 + 6x^3 + \ldots..(2)$

(1) – (2) gives
$S\,(1 \quad x) = 2 + 2x + 2x^2 + 2x^3 + \ldots$
$= 2\,(1 + x + x^2 + \ldots)$

$1 + x + x^2 + \ldots$ is an infinite G.P with $a = 1$, $r = x$ and $|r| = |x| < 1$

$\therefore$ Sum of the series $= 1/1 \;\; x$

$\therefore S\,(1 \quad x) = 2/(1 \quad x)$

$\therefore\; S = 2/(1 \quad x)^2$

Example 10:

Find the sum of the series 1, 2/5, 4/25, 8/125, ... ∏.

Solution:

Given that, $a = 1$, $r = \frac{2}{5}$ and $|r| = \left|\frac{2}{5}\right| < 1$

$$\therefore S_{\prod} = \frac{a}{1 - r} = \frac{1}{1 - \frac{2}{5}} = 5/3$$

Note:

When n geometric means are there between a and b, the common ratio of the G.P. can be derived as follows.

Given that, n geometric means are there between a and b.

$\therefore$ $a = t_1$ and $b = t_{n+2}$

Let 'r' be the common ratio

$b = (t_1)(r^{n+1})$ $\quad b = a\,r^{n+1}$

$$r^{n+1} = \frac{b}{a}$$

$$r = \sqrt[(n+1)]{\frac{b}{a}}$$

Harmonic Progression (H.P.):

A progression is said to be a harmonic progression if the reciprocal of the terms in the progression form an arithmetic progression.
For example, consider the series

$$\frac{1}{2}, \frac{1}{5}, \frac{1}{8}, \frac{1}{11}, \ldots$$

The progression formed by taking reciprocals of terms of the above series is 2, 5, 8, 11,. . . . Clearly, these terms form an A.P. whose common difference is 3.

n^{th} term of an H.P:

We know that if a, a + d, a + 2d,….are in A.P., then the nth term of this A.P. is $a + (n - 1)d$. Its reciprocal is

$$\frac{1}{a+(n-1)d}$$

So, n^{th} term of an H.P. whose first two terms are $\frac{1}{a}$ and $\frac{1}{a+d}$ is $\frac{1}{a+(n-1)d}$

Note:

There is no concise general formula for the sum to n terms of an H.P.

Example 11:

Find the 10th term of the H.P. $\frac{3}{2}, 1, \frac{3}{4}, \frac{3}{5}, \ldots\ldots\ldots\ldots$

Solution:

The given H.P. is $\frac{3}{2}, 1, \frac{3}{4}, \frac{3}{5}, \ldots\ldots$

The corresponding A.P. is $\frac{2}{3}, 1, \frac{4}{3}, \frac{5}{3}, \ldots\ldots$

Here $a = \frac{2}{3}$; $d = 1 - \frac{2}{3} = \frac{1}{3}$

$\therefore$ T_{10} of the corresponding A.P. is $a + (10 - 1)d = \frac{2}{3} + (9)\frac{1}{3} = \frac{11}{3}$

Hence required term in H.P. is $\frac{3}{11}$

Harmonic Mean (H.M.):

If three terms are in H.P., then the middle term is the H.M. of other two terms.

The harmonic mean of two terms a and b is given by

$$\text{H.M.} = \frac{2ab}{a+b}$$

Relation between A.M., H.M. and G.M. of two numbers

Let x and y be two numbers

$\therefore$ A.M. = $\frac{x+y}{2}$, G.M. = $\sqrt{xy}$ and H.M. = $\frac{2xy}{x+y}$

(A.M.) (H.M.) = $(G.M)^2$

Inserting n harmonic means between two numbers:

To insert n H.M.'s between two numbers, we first take the corresponding arithmetic series and insert n arithmetic means, and next, we find the corresponding harmonic series.

This is illustrated by the example below:

Example 12:

Insert three harmonic means between $\frac{1}{12}$ and $\frac{1}{20}$

Solution:

After inserting the harmonic means
let the harmonic progression be

$\frac{1}{a}, \frac{1}{a+d}, \frac{1}{a+2d}, \frac{1}{a+3d}, \frac{1}{a+4d}$

As $\frac{1}{a} = \frac{1}{12}$ and $\frac{1}{a+4d} = \frac{1}{20}$ a = 12 and d = 2

$\therefore$ The required harmonic means are $\frac{1}{14}, \frac{1}{16}$ and $\frac{1}{18}$

Algebra Notes

1. Quadratic formula

$$x = \frac{-b \pm \sqrt{b^2 - 4ac}}{2a}$$

2. Algebraic Identities

$$(a + b)^2 = a^2 + 2ab + b^2$$

$$(a - b)^2 = a^2 - 2ab + b^2$$

$$a^2 - b^2 = (a + b)(a - b)$$

$$a^2 + b^2 = (a + b)^2 - 2ab$$

$$a^2 + b^2 = (a - b)^2 + 2ab$$

3. Distance between 2 points

$P(x_1, y_1)$ $Q(x_2, y_2)$

Then $PQ = \sqrt{(x_2 - y_2)^2 + (x_1 - y_1)^2}$

4. Equation of a Straight line

$y = mx + c$

Where m = slope

c = y - intercept

5. Point slope form of a straight line

$y - y_1 = m(x - x_1)$

Where m is the slope

Straight line passes through the point (x_1, y_1)

6. Slope of a straight

Given 2 points (x_1, y_1) and (x_2, y_2)

$$\text{slope} = \frac{(y_2 - y_1)}{(x_2 - x_1)}$$

$$= \frac{(y_1 - y_2)}{(x_1 - x_2)}$$

Abbreviations

milligram	mg
gram	g
kilogram	kg
milliliter	mL
liter	L
kiloliter	kL
millimeter	mm
centimeter	cm
meter	m
kilometer	km
square centimeter	cm^2
cubic centimeter	cm^3

year	yr
month	mon
hour	hr
minute	min
second	sec

volume	V
total Square Area	S.A
area of base	B
ounce	oz
pound	lb
quart	qt
gallon	gal.
inches	in.
foot	ft
yard	yd
mile	mi.
square inch	sq in.
square foot	sq ft
cubic inch	cu in.
cubic foot	cu ft

Basic Practice

Write the prime factorization of each number given below.

1. 36

(A) $5 \cdot 7$ (B) $2 \cdot 2$ (C) $2 \cdot 2 \cdot 3 \cdot 3$ (D) $3 \cdot 7$

2. 20

(A) $2 \cdot 2 \cdot 2 \cdot 2$ (B) $2 \cdot 7$ (C) $2 \cdot 2 \cdot 5$ (D) $3 \cdot 7$

3. -37

(A) -37 (B) $-2 \cdot 2 \cdot 2 \cdot 3 \cdot 37$ (C) $-2 \cdot 2 \cdot 37$ (D) $-2 \cdot 2 \cdot 2 \cdot 3 \cdot 3 \cdot 37$

4. -71

(A) -71 (B) $-2 \cdot 2 \cdot 2 \cdot 3 \cdot 3 \cdot 71$ (C) $-2 \cdot 2 \cdot 2 \cdot 3 \cdot 71$ (D) $-2 \cdot 2 \cdot 2 \cdot 71$

5. -35

(A) $-5 \cdot 7$ (B) $-2 \cdot 5 \cdot 5 \cdot 7$ (C) $-2 \cdot 2 \cdot 2 \cdot 3 \cdot 5 \cdot 5 \cdot 7$ (D) $-2 \cdot 2 \cdot 5 \cdot 5 \cdot 7$

6. 38

(A) $2 \cdot 3 \cdot 3$ (B) $2 \cdot 3$ (C) $2 \cdot 5$ (D) $2 \cdot 19$

7. -66

(A) $-2 \cdot 3 \cdot 11$ (B) $-2 \cdot 3 \cdot 3 \cdot 5 \cdot 11$ (C) $-2 \cdot 2 \cdot 3 \cdot 3 \cdot 3 \cdot 5 \cdot 11$ (D) $-2 \cdot 3 \cdot 3 \cdot 11$

8. 80

(A) $2 \cdot 3 \cdot 3 \cdot 5$ (B) $3 \cdot 3 \cdot 3 \cdot 3$ (C) $2 \cdot 3 \cdot 23$ (D) $2 \cdot 2 \cdot 2 \cdot 2 \cdot 5$

Evaluate the below numerical expressions.

1) $((41)(5)-(-19)) \div 14$

A) 35 B) -16
C) 16 D) 23

2) $11-(9)(4^2)$

A) −133 B) −150
C) −153 D) −134

3) $((24-7) \div (-17))(-20)$

A) 5 B) 0
C) 31 D) 20

4) $((-6)+1-15)(10)$

A) −204 B) 200
C) −200 D) −193

5) $(-11) \div (3-14) - 17$

A) −16 B) 2
C) −35 D) 16

6) $(-9)-10+8-4$

A) −15 B) −10
C) −17 D) 16

7) $3+(-20)+3^2$

A) −8 B) −26
C) −24 D) 28

8) $(((-56) \div (-8)) - (-16))(-2)$

A) −46 B) −55
C) −66 D) 46

Simplify the below expressions.

1. $-3(1+3b)-6(b-5)$

(A) 27 - 15b (B) 27 - 2b (C) -9 + 35b (D) -150b + 6

2. $-13(-5-5n)-5(6n-4)$

(A) -33 - 99n (B) -11n - 58 (C) 85 + 44n (D) 85 + 35n

3. $-4(2+5x)-6(x-2)$

(A) 4 - 26x (B) 4 - 37x (C) 22 - 107x (D) 38x + 115

4. $-11(x+12)-13(2x-4)$

(A) -37x - 80 (B) -6x + 52 (C) -16x + 52 (D) 7 - 70x

5. $-5(r+9)-12(-4-11r)$

(A) 106r + 3 (B) 103r + 3 (C) 127r + 3 (D) 114r + 3

6. $-3(1-6p)-6(12-10p)$

(A) -86p - 99 (B) -75 + 80 p (C) -75 + 78 p (D) -78 + 24p

7. $-11(1+7v)-4(1-14v)$

(A) -15 - 21v (B) -31v + 124 (C) -18v - 6 (D) -12v - 6

8. $-9(b-11)-(13b+14)$

(A) -22b + 85 (B) -132b - 139 (C) -123b - 139 (D) -111b - 139

Evaluate the below expressions.

1. $7 + p - m$; where $m = -8$, $p = 7$

(A) 25 (B) 26 (C) 22 (D) 20

2. $c + a - c$; where $a = -4$, and $c = -10$

(A) -4 (B) -2 (C) -13 (D) 4

3. $|n| + m$; where $m = 9$, and $n = -1$

(A) 8 (B) 16 (C) 10 (D) 17

4. $\frac{z}{4} + x$; where $x = -8$, and $z = -8$

(A) -10 (B) 0 (C) -6 (D) -2

5. $y(z + y)$; where $y = 2$, and $z = 6$

(A) 24 (B) 16 (C) 12 (D) 15

6. xy^3; where $x = -1$, and $y = 4$

(A) -64 (B) -70 (C) -66 (D) -72

7. $(x + y)^2$; where $x = 7$, and $y = -3$

(A) 26 (B) 22 (C) 6 (D) 16

8. $-3m + n$; where $m = -3$, and $n = 10$

(A) 19 (B) 24 (C) 26 (D) 14

Simplify the below with positive exponents.

1) $\dfrac{2x^5y^3}{(x^2)^0}$

A) $\dfrac{2}{x^{25}y^{16}}$ B) $16x^{19}y^{16}$

C) $2x^5y^3$ D) $\dfrac{8}{y^6}$

2) $\dfrac{(2xy)^2}{2y^4}$

A) $\dfrac{1}{4x^2y}$ B) $\dfrac{1}{32x^3y^{18}}$

C) $\dfrac{y}{4x}$ D) $\dfrac{2x^2}{y^2}$

3) $\dfrac{(2m^5)^5}{m^4}$

A) $\dfrac{1}{4m^{10}n^4}$ B) $32m^{21}$

C) $\dfrac{n^{20}}{m^{12}}$ D) $\dfrac{m^{16}}{2}$

4) $\dfrac{y^4}{(2x^2y^3)^5}$

A) $\dfrac{x^{11}y^8}{2}$ B) $\dfrac{1}{4x^7y^4}$

C) $\dfrac{1}{2x^2y^5}$ D) $\dfrac{1}{32x^{10}y^{11}}$

5) $\dfrac{(2x^4y^3)^4}{x^3y^3}$

A) $\dfrac{1}{yx^5}$ B) $16x^{13}y^9$

C) $\dfrac{1}{x^4y^4}$ D) $\dfrac{4y^{12}}{x^3}$

6) $\dfrac{x^3}{(x^4y^5)^4}$

A) $x^{20}y^9$ B) $\dfrac{x}{y^7}$

C) $\dfrac{1}{x^{13}y^{20}}$ D) $16y^2x^7$

7) $\dfrac{(2vu^4)^2}{u}$

A) $16v^4u^6$ B) $\dfrac{1}{32u^{19}}$

C) $4v^2u^7$ D) $\dfrac{8}{u^3}$

8) $\dfrac{x^3y^4}{(2x^3y^2)^2}$

A) $\dfrac{1}{4x^8y^3}$ B) $16x^{24}y^{16}$

C) $\dfrac{1}{2x^2y^4}$ D) $\dfrac{1}{4x^3}$

Find the distance between the below points.

1. A(4, 6), B(−4, −8)

(A) $2\sqrt{65}$ (B) $\sqrt{6}$ (C) $\sqrt{22}$ (D) 2

2. C(−2, 5), D(−4, −2)

(A) $3\sqrt{5}$ (B) $\sqrt{53}$ (C) $3\sqrt{3}$ (D) 3

3. E(4, 5), F(8, 3)

(A) $2\sqrt{5}$ (B) $4\sqrt{5}$ (C) $4\sqrt{13}$ (D) $\sqrt{6}$

4. G(−7, 1), H(−6, −1)

(A) 13 (B) $\sqrt{5}$ (C) $\sqrt{13}$ (D) $\sqrt{3}$

5. I(−7, −7), J(1, -1)

(A) $\sqrt{22}$ (B) 10 (C) $\sqrt{10}$ (D) 5

6. K(−8, −7), L(−7, −1)

(A) 17 (B) $\sqrt{7}$ (C) $2\sqrt{2}$ (D) $\sqrt{37}$

7. M(4, 4), N(6, −6)

(A) $2\sqrt{22}$ (B) $4\sqrt{6}$ (C) $2\sqrt{26}$ (D) $\sqrt{51}$

Find the midpoint of each line segment given in below graphs.

1.

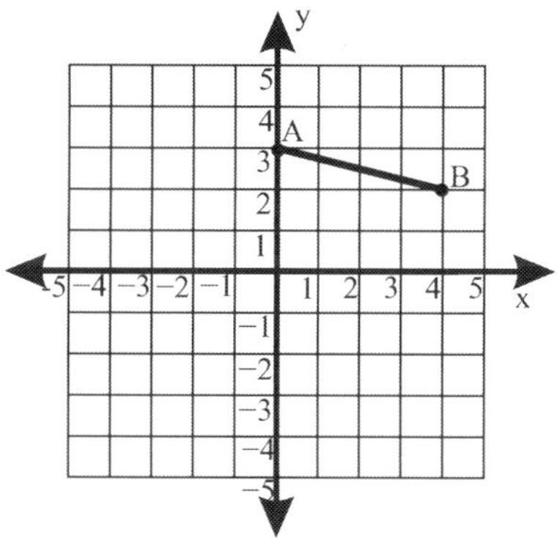

(A) (1.5, 3) (B) (−2, 0.5) (C) (8, 1) (D) (2, 1.5)

2.

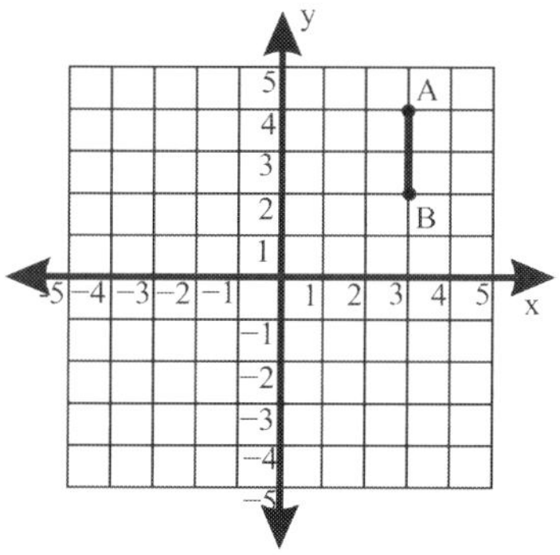

(A) (3.5, 2.5) (B) (0, 1) (C) (3, 0) (D) (3, 3)

3.

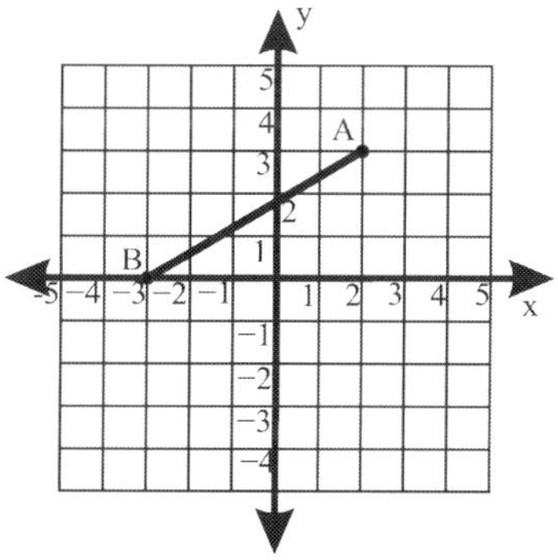

(A) (−2.5, −1.5) (B) (−0.5, 1.5) (C) (−1.5, 2.5) (D) (7, 6)

Find the midpoint of each line segment given in below graphs.

4.

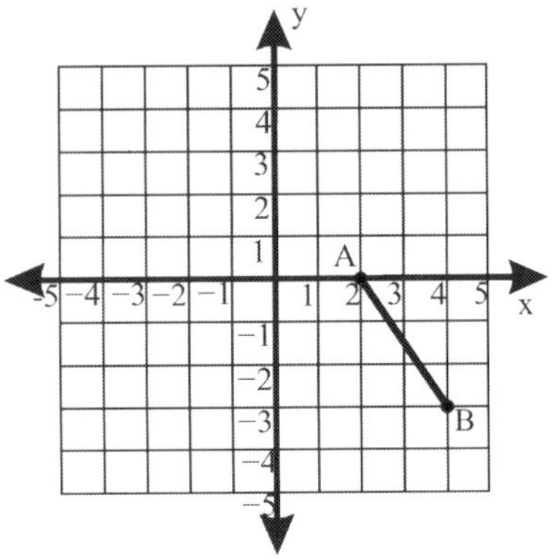

(A) (3, −1.5) (B) (1, 0.5) (C) (6, −6) (D) (−1, 1.5)

5.

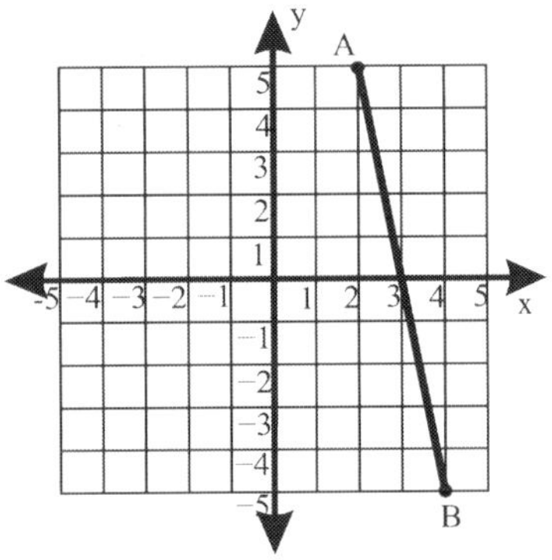

(A) (3, 0) (B) (−0.5, 3.5) (C) (0, 15) (D) (1, −5)

6.

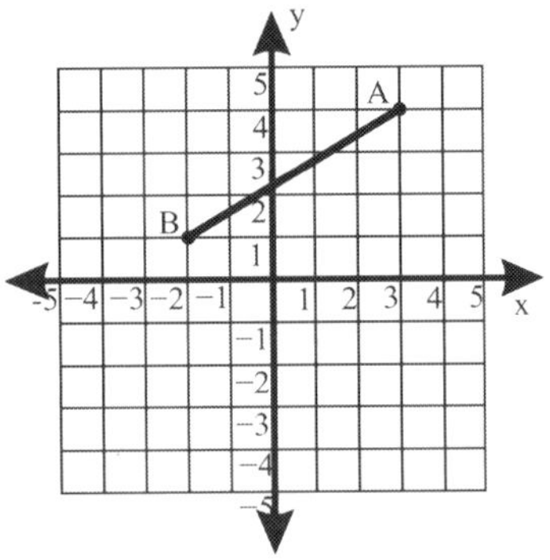

(A) (−2.5, −1.5) (B) (−0.5, 3.5) (C) (8, 7) (D) (0.5, 2.5)

Find the midpoint of each line segment given in below graphs.

7.

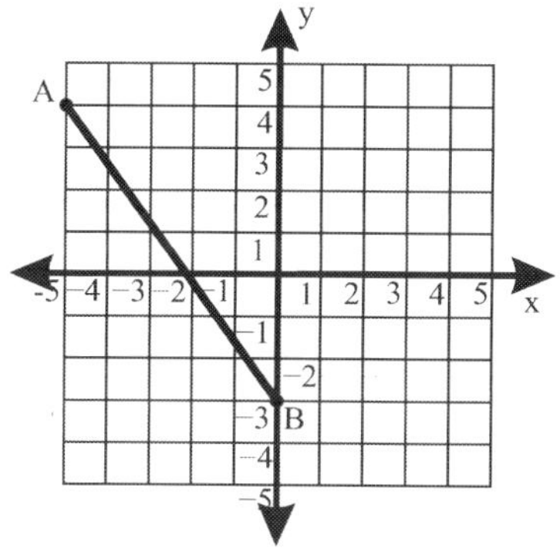

(A) (−1.5, −0.5) (B) (−10, 11) (C) (−2.5, 0.5) (D) (2.5, −3.5)

8.

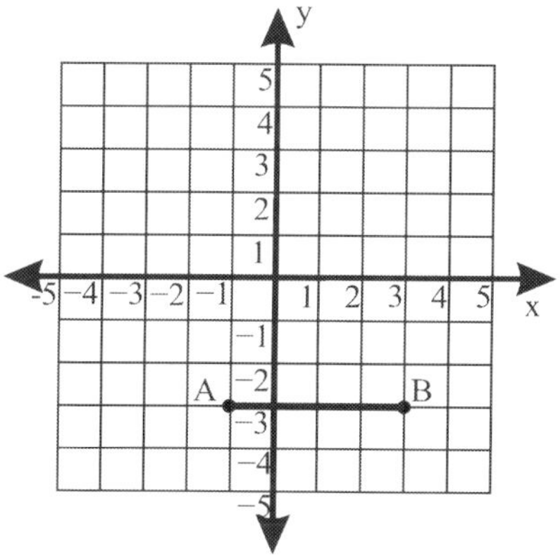

(A) (7, −3) (B) (0, 1) (C) (−2, 0) (D) (1, −3)

Write the standard form of the straight line passing through the given points below.

1. A(2, 4) and B(0, −3)

(A) $2x + 5y = 7$ (B) $10x + 2y = -7$ (C) $7x - 2y = 6$ (D) $2x + 7y = 10$

2. C(0, 2) and D(1, 4)

(A) $2x + 2y = -1$ (B) $2x - y = -2$ (C) $4x + y = -1$ (D) $5x - y = -1$

3. E(2, −5) and F(0, −5)

(A) $5x = 1$ (B) $x + 2y = 0$ (C) $y = -5$ (D) $2x = 1$

4. G(3, −1) and H(−2, 0)

(A) $x - 5y = -2$ (B) $5x - y = 2$ (C) $x + y = 3$ (D) $x + 5y = -2$

5. I(3, −4) and J(0, −1)

(A) $4x + 3y = -1$ (B) $x + y = -1$ (C) $3x + 3y = 1$ (D) $3x - y = 4$

6. K(0, 2) and L(5, −2)

(A) $4x - 5y = -4$ (B) $5x - 4y = 10$ (C) $4x + 5y = -4$ (D) $4x + 5y = 10$

7. M(0, 4) and N(2, −2)

(A) $3x + y = 4$ (B) $x - y = -5$ (C) $x - y = 2$ (D) $x + y = 2$

Find the slope of the straight line passing through the points given below.

1. A(18, −15), B(18, −17)

(A) Undefined

(B) −1

(C) 1

(D) 0

2. M(−9, − 14), N(−9, 7)

(A) $-\frac{5}{2}$

(B) Undefined

(C) 0

(D) $\frac{5}{2}$

3. O(−1, 1), P(−15, −1)

(A) $-\frac{1}{7}$

(B) $\frac{1}{7}$

(C) −7

(D) 7

4. Q(−1, 9), R(9, −6)

(A) $\frac{2}{3}$

(B) $-\frac{2}{3}$

(C) $-\frac{3}{2}$

(D) $\frac{3}{2}$

5. S(−17, 20), T(−16, −17)

(A) −37

(B) $-\frac{1}{37}$

(C) $\frac{1}{37}$

(D) 37

6. U(−6, 12), V(−6, 8)

(A) −1

(B) 1

(C) Undefined

(D) 0

Find the slope-intercept form of the straight line with the given slope and y-intercept for all the problems given below.

1. Slope $= -\frac{2}{5}$, y-intercept = 5

(A) $y = -5x - \frac{2}{5}$

(B) $y = \quad x + 5 - \frac{2}{5}$

(C) $y = 5x - \frac{2}{5}$

(D) $y = -\frac{2}{5}x - 5$

2. Slope $= \frac{7}{4}$, y-intercept = 5

(A) $y = x + 5$

(B) $y = 5x + 1$

(C) $y = \frac{7}{4}x + 5$

(D) $y = -x + 1$

3. Slope $= -\frac{9}{4}$, y-intercept = −4

(A) $y = -4x + \frac{1}{2}$

(B) $y = -\frac{9}{4}x - 4$

(C) $y = -\frac{1}{2}x - 4$

(D) $y = \frac{1}{2}x - 4$

4. Slope = −3, y-intercept = 5

(A) $y = -3x + 5$

(B) $y = 5x - 3$

(C) $y = -5x - 3$

(D) $y = -3x - 5$

Find the slope-intercept form of the straight line with the given slope and y-intercept for all the problems given below.

5. Slope = 6, y-intercept = 3

(A) $y = -5x - 6$

(B) $y = 3x - 6$

(C) $y = -6x + 3$

(D) $y = 6x + 3$

6. Slope = $\frac{3}{5}$, y-intercept = −4

(A) $y = \frac{3}{5}x + 4$

(B) $y = 4x + \frac{3}{5}$

(C) $y = \frac{3}{5}x - 4$

(D) $y = -4x + \frac{3}{5}$

7. Slope = $\frac{1}{2}$, y-intercept = −3

(A) $y = -3x - \frac{1}{2}$

(B) $y = -2x - 3$

(C) $y = x - 3 - \frac{1}{2}$

(D) $y = \frac{1}{2}x - 3$

Find the slope of the straight line for each of the questions below.

1)

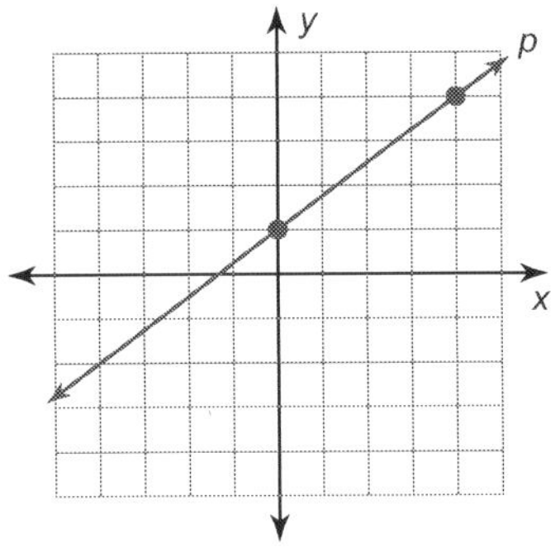

A) $\frac{4}{3}$ B) $\frac{3}{4}$

C) $-\frac{4}{3}$ D) $-\frac{3}{4}$

2)

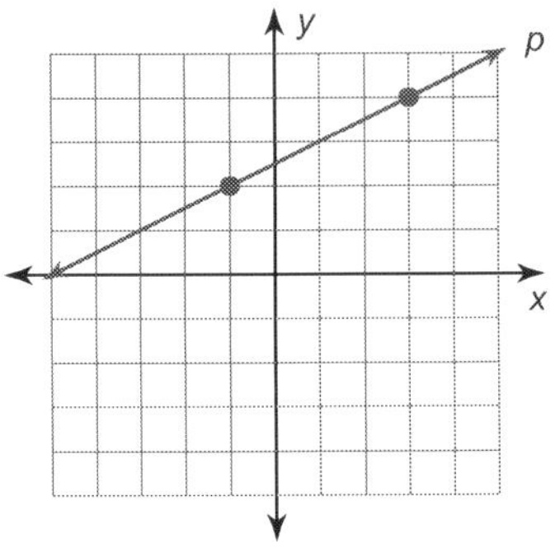

A) 2 B) $\frac{1}{2}$

C) −2 D) $-\frac{1}{2}$

3)

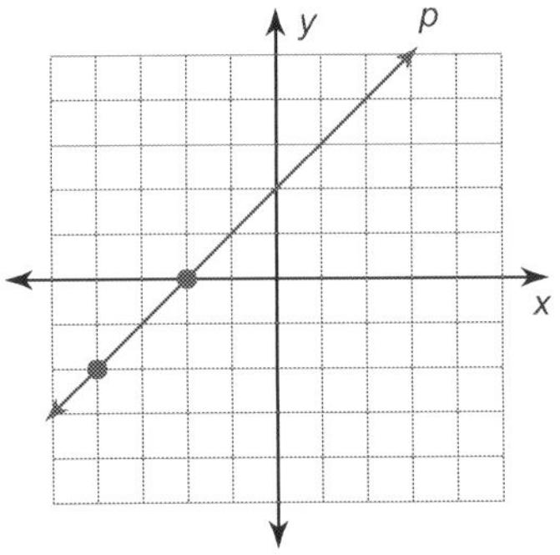

A) $-\frac{4}{5}$ B) 1

C) −1 D) $\frac{4}{5}$

4)

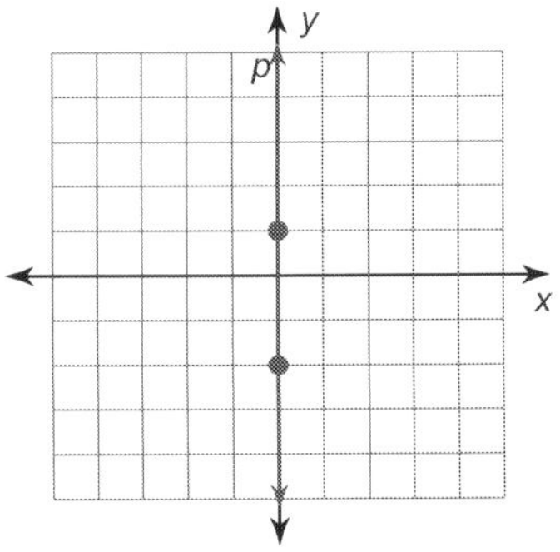

A) $\frac{3}{4}$ B) $-\frac{3}{4}$

C) 0 D) Undefined

Find the slope of the straight line for each of the questions below.

5)

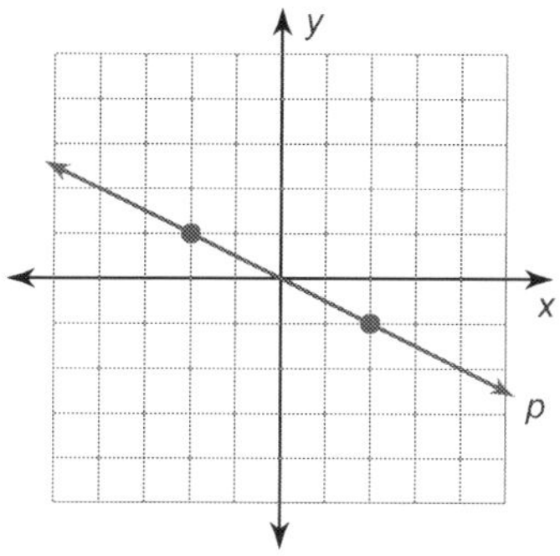

A) $\frac{1}{2}$ B) $-\frac{1}{2}$

C) -2 D) 2

6)

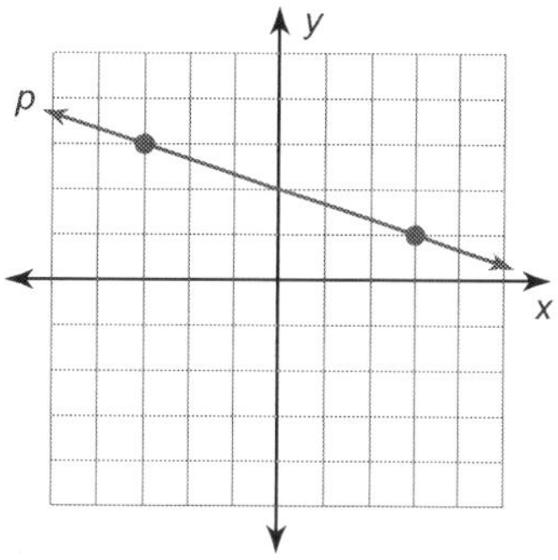

A) 3 B) $\frac{1}{3}$

C) -3 D) $-\frac{1}{3}$

7)

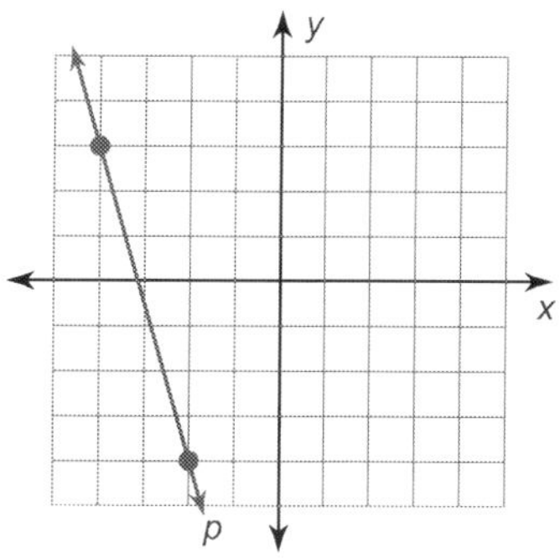

A) $\frac{7}{2}$ B) $-\frac{7}{2}$

C) $-\frac{2}{7}$ D) $\frac{2}{7}$

8)

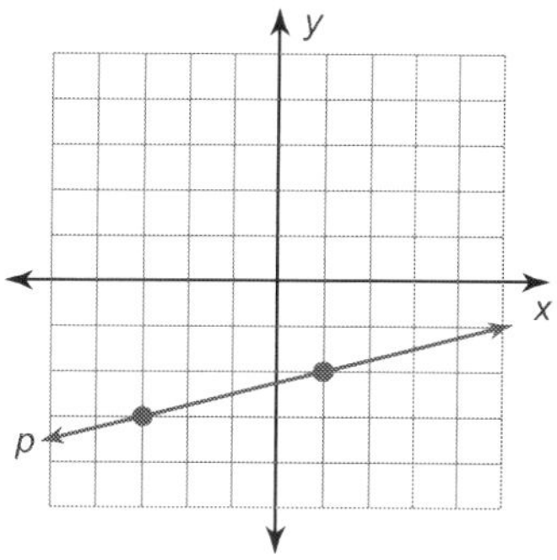

A) $\frac{1}{4}$ B) -4

C) 4 D) $-\frac{1}{4}$

Find the slope of the straight line for each of the questions below.

1) $x + y = -3$

A) -1 B) 1 C) $\frac{3}{2}$ D) $-\frac{3}{2}$

2) $4x + 3y = 3$

A) $-\frac{4}{3}$ B) $-\frac{3}{4}$ C) $\frac{3}{4}$ D) $\frac{4}{3}$

3) $2x - 3y = 6$

A) $\frac{2}{3}$ B) $-\frac{2}{3}$ C) $\frac{3}{2}$ D) $-\frac{3}{2}$

4) $3x - 5y = -10$

A) $\frac{5}{3}$ B) $-\frac{5}{3}$ C) $-\frac{3}{5}$ D) $\frac{3}{5}$

5) $5x + 2y = 0$

A) $-\frac{5}{2}$ B) $-\frac{2}{5}$ C) $\frac{5}{2}$ D) $\frac{2}{5}$

6) $4x + y = 1$

A) 4 B) $\frac{1}{4}$ C) $-\frac{1}{4}$ D) -4

7) $x + 2y = 6$

A) 2 B) $\frac{1}{2}$ C) $-\frac{1}{2}$ D) -2

8) $5x - 3y = -12$

A) $-\frac{5}{3}$ B) $\frac{3}{5}$ C) $\frac{5}{3}$ D) $-\frac{3}{5}$

Find the slope of a line parallel to each given line.

1) $y = -5$

A) -2 B) Undefined
C) 0 D) 2

2) $5x - 4y = 0$

A) $\frac{5}{4}$ B) $\frac{4}{5}$
C) $-\frac{5}{4}$ D) $-\frac{4}{5}$

3) $x = 1$

A) Undefined B) $-\frac{1}{2}$
C) 0 D) $\frac{1}{2}$

4) $x + 5y = 0$

A) $\frac{1}{5}$ B) 5
C) $-\frac{1}{5}$ D) -5

5) $x - 4y = -20$

A) -4 B) 4
C) $\frac{1}{4}$ D) $-\frac{1}{4}$

6) $x + 4y = -16$

A) 4 B) $-\frac{1}{4}$
C) $\frac{1}{4}$ D) -4

7) $5x - 3y = -9$

A) $-\frac{3}{5}$ B) $-\frac{5}{3}$
C) $\frac{5}{3}$ D) $\frac{3}{5}$

8) $3x + 2y = 4$

A) $-\frac{3}{2}$ B) $-\frac{2}{3}$
C) $\frac{3}{2}$ D) $\frac{2}{3}$

Find the slope of a line perpendicular to each given line.

1) $y = -3x + 2$

A) $\frac{1}{3}$　B) 3

C) -3　D) $-\frac{1}{3}$

2) $y = -\frac{4}{5}x - 3$

A) $\frac{4}{5}$　B) $\frac{5}{4}$

C) $-\frac{5}{4}$　D) $-\frac{4}{5}$

3) $y = -4x + 4$

A) -4　B) $\frac{1}{4}$

C) $-\frac{1}{4}$　D) 4

4) $y = -\frac{3}{2}x - 4$

A) $\frac{3}{2}$　B) $-\frac{2}{3}$

C) $-\frac{3}{2}$　D) $\frac{2}{3}$

5) $y = -\frac{5}{2}x + 2$

A) $\frac{5}{2}$　B) $\frac{2}{5}$

C) $-\frac{5}{2}$　D) $-\frac{2}{5}$

6) $y = x - 1$

A) -1　B) 1

C) $-\frac{2}{3}$　D) $\frac{2}{3}$

7) $y = \frac{6}{5}x + 1$

A) $-\frac{5}{6}$　B) $-\frac{6}{5}$

C) $\frac{6}{5}$　D) $\frac{5}{6}$

8) $y = 3x + 4$

A) $-\frac{1}{3}$　B) $\frac{1}{3}$

C) 3　D) -3

Simplify the below radicals.

1. $\dfrac{-7-\sqrt{2}}{7+\sqrt{6}}$

(A) $\dfrac{-35+7\sqrt{6}-5\sqrt{2}+2\sqrt{3}}{19}$

(B) $\dfrac{-49+7\sqrt{6}+14\sqrt{2}-4\sqrt{3}}{43}$

(C) $\dfrac{-49+7\sqrt{6}-7\sqrt{2}+2\sqrt{3}}{43}$

(D) $\dfrac{-63+7\sqrt{6}-9\sqrt{2}-2\sqrt{3}}{75}$

2. $\dfrac{5-\sqrt{5}}{3-6\sqrt{2}}$

(A) $\dfrac{-5-10\sqrt{2}+\sqrt{5}+2\sqrt{10}}{21}$

(B) $\dfrac{-5-120\sqrt{2}+2\sqrt{5}+48\sqrt{10}}{1151}$

(C) $\dfrac{-5+10\sqrt{2}-\sqrt{5}+2\sqrt{10}}{21}$

(D) $\dfrac{-2+10\sqrt{2}+\sqrt{5}+2\sqrt{10}}{27}$

3. $\dfrac{-10-\sqrt{10}}{10+\sqrt{6}}$

(A) $\dfrac{-60+5\sqrt{6}-6\sqrt{10}+\sqrt{15}}{69}$

(B) $\dfrac{-100+10\sqrt{6}-10\sqrt{13}+\sqrt{78}}{94}$

(C) $\dfrac{-50+5\sqrt{6}-10\sqrt{3}+3\sqrt{2}}{47}$

(D) $\dfrac{-50+5\sqrt{6}-5\sqrt{10}+\sqrt{15}}{47}$

4. $\dfrac{10\sqrt{5}-2\sqrt{7}}{2-2\sqrt{3}}$

(A) $\dfrac{-12\sqrt{5}+8\sqrt{7}}{5}$

(B) $\dfrac{-5\sqrt{5}-5\sqrt{15}+\sqrt{7}+\sqrt{21}}{2}$

(C) $\dfrac{-5\sqrt{5}-5\sqrt{15}}{2}$

(D) $\dfrac{-6\sqrt{5}+6\sqrt{15}-\sqrt{7}-\sqrt{21}}{75}$

Simplify the below radicals.

5. $\dfrac{-4+\sqrt{7}}{4\sqrt{5}-\sqrt{6}}$

(A) $\dfrac{-28\sqrt{5}-4\sqrt{6}+7\sqrt{35}+\sqrt{42}}{47}$

(B) $\dfrac{-16\sqrt{5}-4\sqrt{6}+4\sqrt{35}+\sqrt{42}}{74}$

(C) $\dfrac{-8\sqrt{5}-2\sqrt{6}+10\sqrt{2}+\sqrt{15}}{37}$

(D) $\dfrac{-11\sqrt{5}-4\sqrt{6}+13\sqrt{33}+\sqrt{43}}{74}$

6. $\dfrac{8}{9\sqrt{3}-5\sqrt{5}}$

(A) $\sqrt{5}$

(B) $\dfrac{36\sqrt{3}+20\sqrt{5}}{59}$

(C) $\dfrac{-48\sqrt{3}-40\sqrt{5}}{17}$

(D) $\dfrac{36\sqrt{13}+20\sqrt{5}}{59}$

7. $\dfrac{7+6\sqrt{10}}{\sqrt{3}+4}$

(A) $\dfrac{-7\sqrt{3}+28-6\sqrt{30}+24\sqrt{10}}{13}$

(B) $\dfrac{-7\sqrt{3}+28-5\sqrt{30}+20\sqrt{10}}{13}$

(C) $\dfrac{-7\sqrt{3}+42-6\sqrt{30}+36\sqrt{10}}{33}$

(D) $\dfrac{-7\sqrt{3}+28-8\sqrt{2}+18\sqrt{10}}{13}$

Simplify the below to the lowest positive terms.

1. $$\frac{\sqrt{40x^2y^2}}{9\sqrt{35x^4y^4}}$$

(A) $\frac{2\sqrt{14}}{77xy}$

(B) $\frac{2\sqrt{14}}{63xy}$

(C) $\frac{2\sqrt{95}}{171xy}$

(D) $\frac{\sqrt{14}}{28xy}$

2. $$\frac{9\sqrt{3x^3y^2}}{9\sqrt{6x^2y^4}}$$

(A) $\frac{\sqrt{30x}}{6y}$

(B) $\frac{7\sqrt{2x}}{18y}$

(C) $\frac{\sqrt{2x}}{2y}$

(D) $\frac{\sqrt{2x}}{3y}$

3. $$\frac{5\sqrt{5p^3}}{\sqrt{7p^3}}$$

(A) $\frac{5\sqrt{35}}{7}$

(B) $\frac{5\sqrt{10}}{4}$

(C) $\frac{5\sqrt{2}}{2}$

(D) $\frac{5\sqrt{2}}{3}$

Simplify the below to the lowest positive terms.

4. $\dfrac{\sqrt{10x^2}}{\sqrt{7x^3}}$

(A) $\dfrac{\sqrt{5x}}{2x}$

(B) $\dfrac{\sqrt{10x}}{3x}$

(C) $\dfrac{\sqrt{70x}}{7x}$

(D) $\dfrac{\sqrt{10x}}{2x}$

5. $\dfrac{8\sqrt{9n}}{\sqrt{2n^2}}$

(A) $\dfrac{15\sqrt{2n}}{n}$

(B) $\dfrac{12\sqrt{2n}}{n}$

(C) $\dfrac{24\sqrt{n}}{n}$

(D) $\dfrac{27\sqrt{2n}}{2n}$

6. $\dfrac{\sqrt{8p}}{7\sqrt{7p^2}}$

(A) $\dfrac{\sqrt{14p}}{28p}$

(B) $\dfrac{2\sqrt{14p}}{49p}$

(C) $\dfrac{\sqrt{70p}}{49p}$

(D) $\dfrac{2\sqrt{3p}}{21p}$

7. $\dfrac{4\sqrt{7a^2b^4}}{\sqrt{8ab}}$

(A) $26\sqrt{6ab}$

(B) $\dfrac{4b\sqrt{35ab}}{5}$

(C) $b\sqrt{14ab}$

(D) $2b\sqrt{3ab}$

Solve the below inequalities.

1. $-7x - 5(x - 1) \leq 89$

(A) No solution (B) $x \geq -10$ (C) $x \geq -7$ (D) $x \geq -6$

2. $123 > -3 + 6(1 - 5x)$

(A) $x > -4$ (B) $x > -23$ (C) $x > -34$ (D) $x > -5$

3. $5(5x - 3) + 8 < -132$

(A) $x < -39$ (B) $x < -10$ (C) $x < -5$ (D) $x < -35$

4. $282 < -6 - 8(-1 + 5k)$

(A) No solution (B) {All real numbers} (C) $k < -37$ (D) $k < -7$

5. $7(x - 7) \leq -84$

(A) $x \leq -12$ (B) $x \leq -5$ (C) $x \leq -40$ (D) $x \leq -28$

6. $87 \leq -4(-6 + 2r) - r$

(A) $r \geq -2$ (B) $r \leq -7$ (C) $r \leq -16$ (D) $r \geq -16$

7. $-84 \geq 7(-6 + 1k)$

(A) $k \leq 2$ (B) No solution (C) $k \leq -6$ (D) $k \leq -27$

8. $-105 < -5(1 - 5p)$

(A) $p > -8$ (B) $p > -22$ (C) $p > -24$ (D) $p > -4$

1. Dan was 70 years old when I visited him 9 years ago. How old is he now ?

   (A) 79 (B) 61 (C) 75 (D) 88

2. How old is Jason if he was 75 years old thirteen years ago ?

   (A) 62 (B) 101 (C) 100 (D) 88

3. Sam spent $42 on soda packs. If they cost $6 / pack, how many packs did he buy ?

   (A) 7 (B) 12 (C) 11 (D) 10

4. If the weight of a package is multiplied by $\frac{3}{4}$ the result is 61.2 lbs. Find the weight of the package.

   (A) 81.6 (B) 61.1 (C) 70.2 (D) 57.4

5. A recipe for pancakes calls for 7 cups of flour. Lucy accidentally put in 10 cups. How many extra cups did she put in ?

   (A) 5 (B) 2 (C) 1 (D) 3

6. Mason and his best friend got a cash prize. They divided the money evenly, each getting $26. How much money was the cash prize ?

   (A) $48 (B) $52 (C) $42 (D) $13

7. Julia and her friends ate 20 muffins. If they ate $\frac{5}{8}$ of what they had originally. How many muffins were there ?

   (A) 36 (B) 32 (C) 12.5 (D) 34

1. Find the area of the circle given diameter = 4 m.

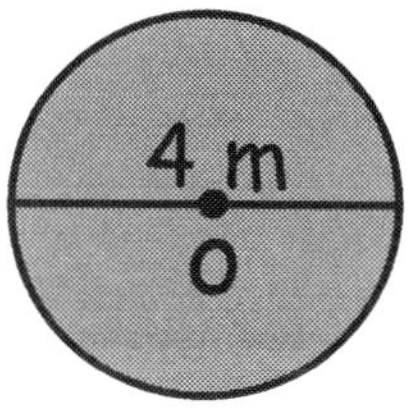

(A) 12∏ m^2

(B) 9∏ m^2

(C) 7∏ m^2

(D) 4∏ m^2

2. Find the area of the circle given diameter = 10 m.

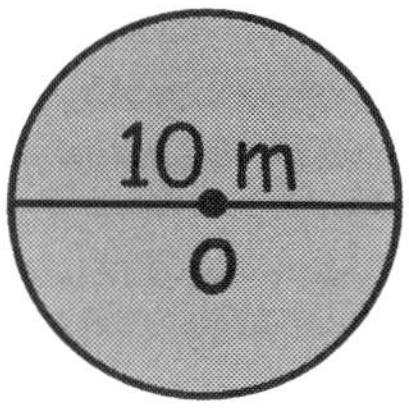

(A) 25∏ m^2

(B) 50∏ m^2

(C) 42∏ m^2

(D) 34∏ m^2

3. Find the area of the circle given diameter = 6 in.

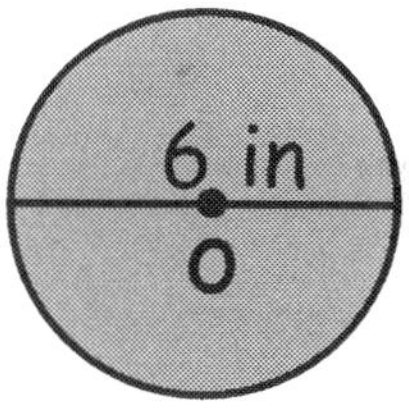

(A) 4∏ in^2

(B) 12∏ in^2

(C) 16∏ in^2

(D) 9∏ in^2

4. Find the area of the circle given diameter = 24 ft.

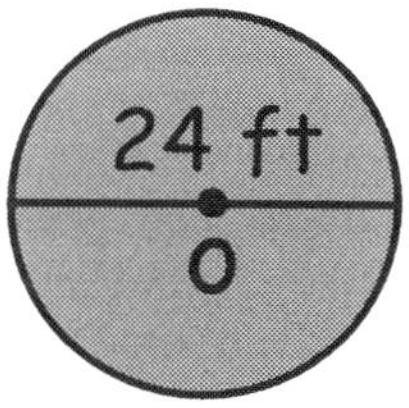

(A) 1296∏ ft^2

(B) 144∏ ft^2

(C) 36∏ ft^2

(D) 9∏ ft^2

5. Find the area of the circle given diameter = 12 in

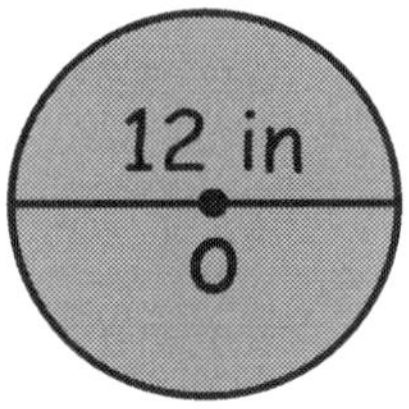

(A) $36\prod$ in^2 (B) $62\prod$ in^2

(C) $44\prod$ in^2 (D) $52\prod$ in^2

6. Find the area of the circle given diameter = 22 yd.

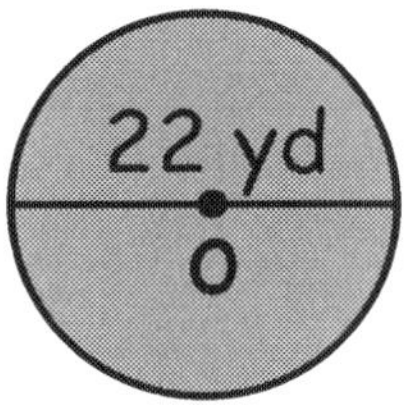

(A) $64\prod$ yd^2 (B) $16\prod$ yd^2

(C) $121\prod$ yd^2 (D) $22\prod$ yd^2

7. Find the area of the circle given diameter = 14 km.

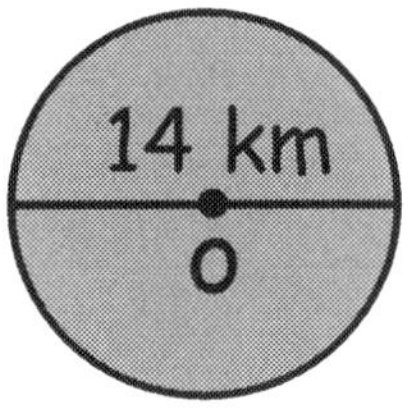

(A) $64\prod$ km^2 (B) $8\prod$ km^2

(C) $49\prod$ km^2 (D) $59\prod$ km^2

8. Find the area of the circle given diameter = 8 ft.

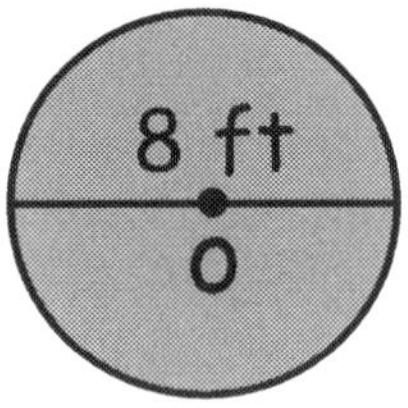

(A) $22\prod$ ft^2 (B) $9\prod$ ft^2

(C) $16\prod$ ft^2 (D) $6\prod$ ft^2

1. Find the volume of the sphere with a radius of 13.2 mi and round the answer to the nearest tenth.

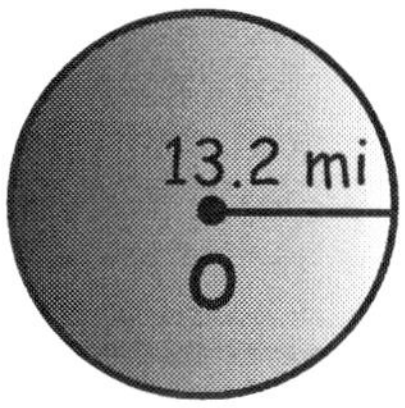

(A) 9634.1 mi^3 (B) 5587.7 mi^3

(C) 6705.2 mi^3 (D) 4817 mi^3

2. Find the volume of the sphere with a diameter of 13.6 mi and round the answer to the nearest tenth.

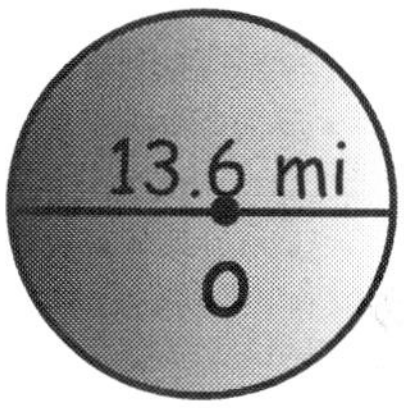

(A) 2370.8 mi^3 (B) 1185.4 mi^3

(C) 2679 mi^3 (D) 1317.1 mi^3

3. Find the volume of the sphere with a diameter of 12.6 m and round the answer to the nearest tenth.

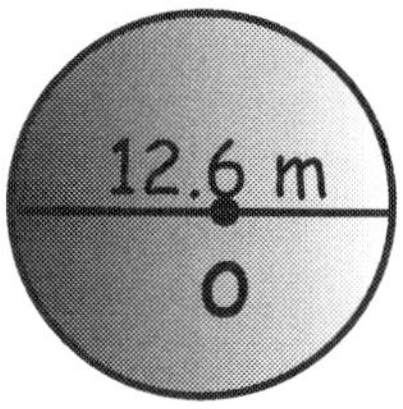

(A) 1149.9 m^3 (B) 1047.4 m^3

(C) 999.9 m^3 (D) 900.8 mi^3

4. Find the volume of the sphere with a radius of 12 ft and round the answer to the nearest tenth.

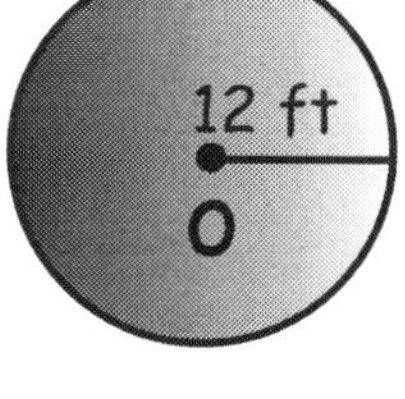

(A) 7238.2 ft^3 (B) 4620.9 ft^3

(C) 3619.1 ft^3 (D) 4125.8 ft^3

5. Find the volume of the sphere with a diameter of 24.6 mi and round the answer to the nearest tenth.

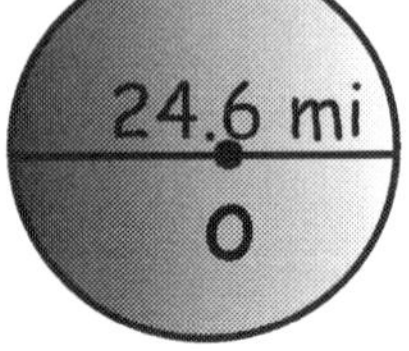

(A) 5827.4 mi^3

(B) 5244.7 mi^3

(C) 7794.8 mi^3

(D) 6547.6 mi^3

6. Find the volume of the sphere with a diameter of 2.2 ft and round the answer to the nearest tenth.

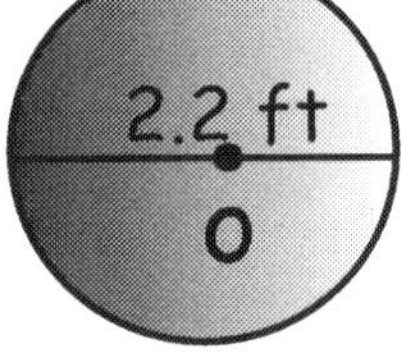

(A) 6.5 ft^3

(B) 13 ft^3

(C) 11.4 ft^3

(D) 5.6 ft^3

7. Find the volume of the sphere with a diameter of 2.1 cm and round the answer to the nearest tenth.

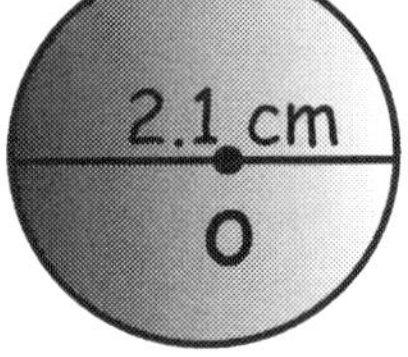

(A) 3.05 cm^3

(B) 3.88 cm^3

(C) 3.5 cm^3

(D) 4.84 cm^3

8. Find the volume of the sphere with a radius of 17.2 cm and round the answer to the nearest tenth.

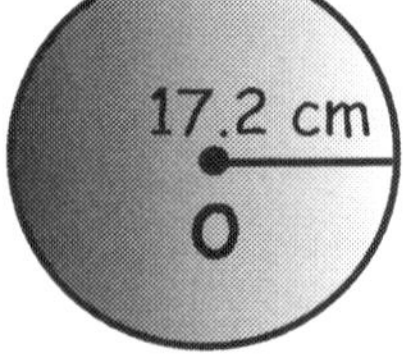

(A) 21314.5 cm^3

(B) 23659.1 cm^3

(C) 42628.9 cm^3

(D) 21314.4 cm^3

1. Find the volume of the rectangular prism measuring 1 ft and 4 ft along the base and 2 ft tall. Round the answer to the nearest tenth.

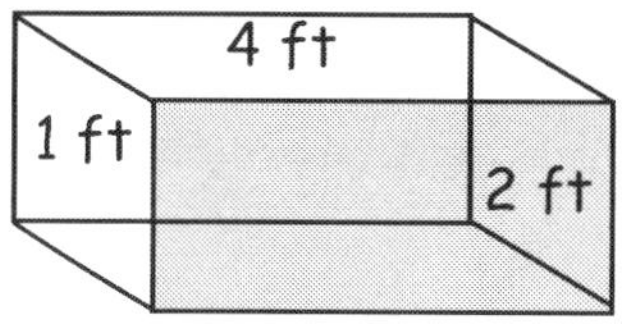

(A) 8 ft^3 (B) 4 ft^3

(C) 9.2 ft^3 (D) 4.6 ft^3

2. Find the volume of the rectangular prism measuring 15 in. and 18 in. along the base and 15 in. tall. Round the answer to the nearest tenth.

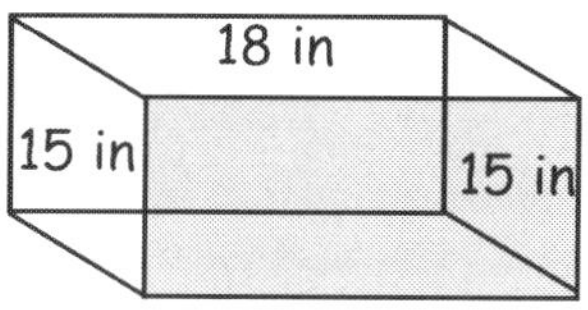

(A) 3361.5 in^3 (B) 2689.2 in^3

(C) 3173.3 in^3 (D) 4050 in^3

3. Find the volume of a square prism measuring 3 yd along each edge of the base and 17 yd tall. Round the answer to the nearest tenth.

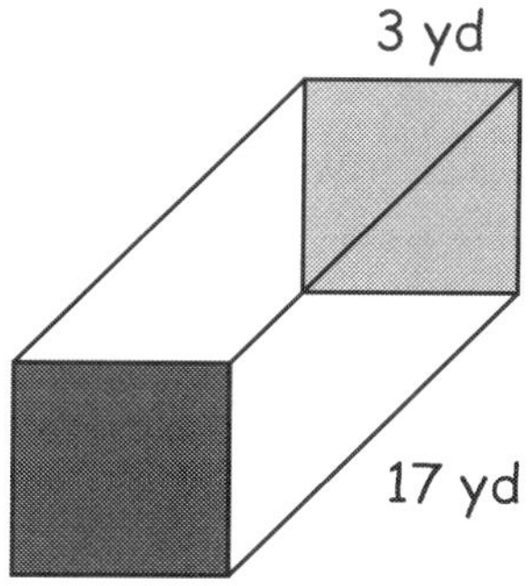

(A) 128.5 yd^3 (B) 153 yd^3

(C) 115.8 yd^3 (D) 131.6 yd^3

4. Find the volume of a rectangular prism measuring 11 ft and 3 ft along the base and 7 ft tall. Round the answer to the nearest tenth.

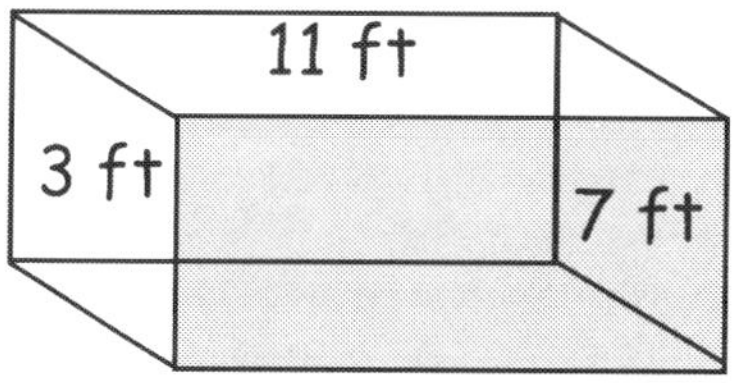

(A) 924 ft^3 (B) 462 ft^3

(C) 1848 ft^3 (D) 231 ft^3

5. Find the volume of a square prism measuring 17 km along each edge of the base and 19 km tall. Round the answer to the nearest tenth.

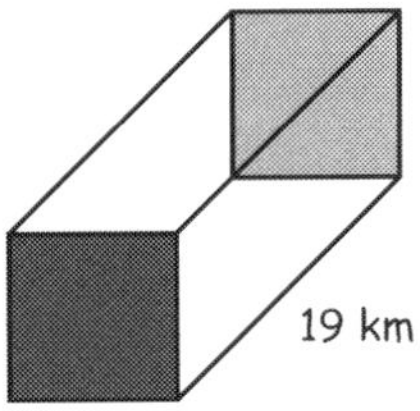

(A) 4101.8 km^3 (B) 5491 km^3

(C) 3363.5 km^3 (D) 4557.5 km^3

6. Find the volume of a square prism measuring 10 cm along each edge of the base and 7 cm tall. Round the answer to the nearest tenth.

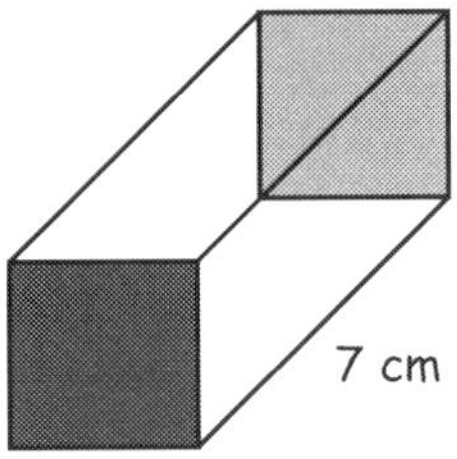

(A) 1416.2 cm^3 (B) 833 cm^3

(C) 708.1 cm^3 (D) 700 cm^3

7. Find the volume of a square prism measuring 12 in. along each edge of the base and 14 in. tall. Round the answer to the nearest tenth.

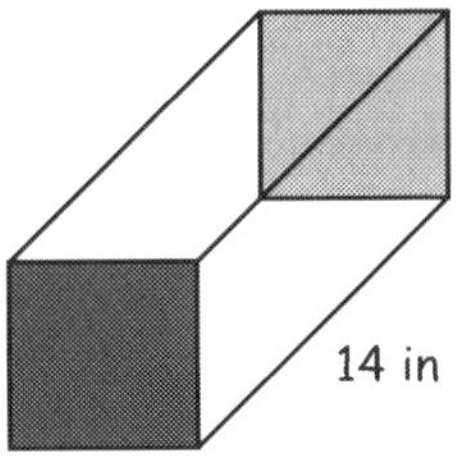

(A) 2016 in^3 (B) 4032 in^3

(C) 4636.8 in^3 (D) 2318.4 in^3

8. Find the volume of a rectangular prism measuring 18 m and 7 m along the base and 16 m tall. Round the answer to the nearest tenth.

(A) 2016 m^3 (B) 4032 m^3

(C) 1633 m^3 (D) 1453.4 m^3

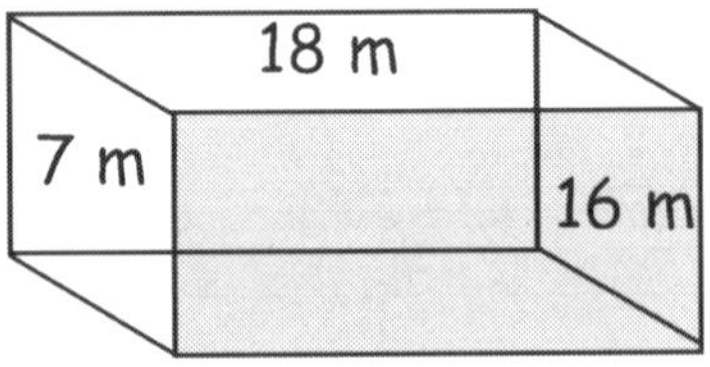

1. Find the volume of a cylinder with a diameter of 34 ft and a height of 12 ft.
Round the answer to the nearest tenth.

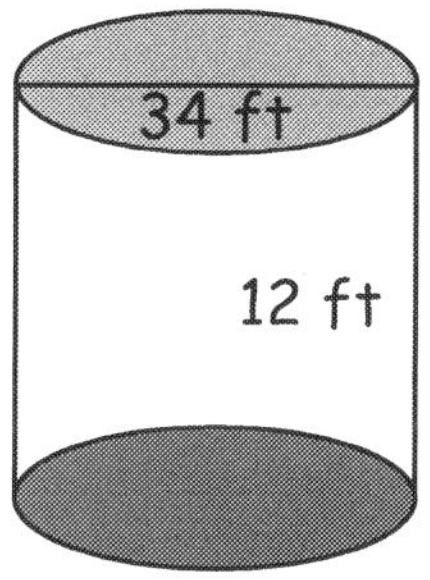

(A) 21598.4 ft^3 (B) 10895 ft^3

(C) 21790.1 ft^3 (D) 18303.7 ft^3

2. Find the volume of a cylinder with a radius of 6 mm and a height of 6 mm.
Round the answer to the nearest tenth.

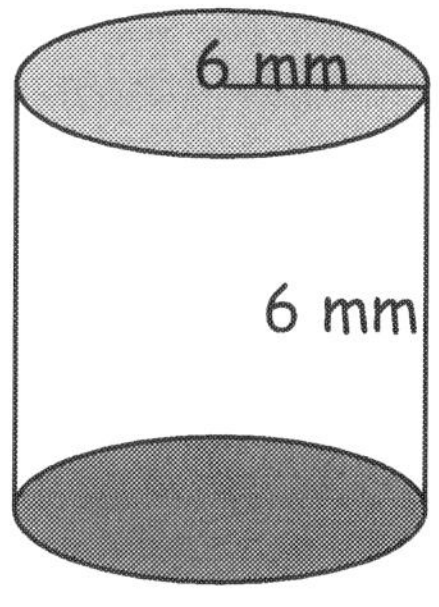

(A) 678.6 mm^3 (B) 3013 mm^3

(C) 1357.2 mm^3 (D) 2714.4 mm^3

3. Find the volume of a cylinder with a diameter of 22 in and a height of 15 in.
Round the answer to the nearest tenth.

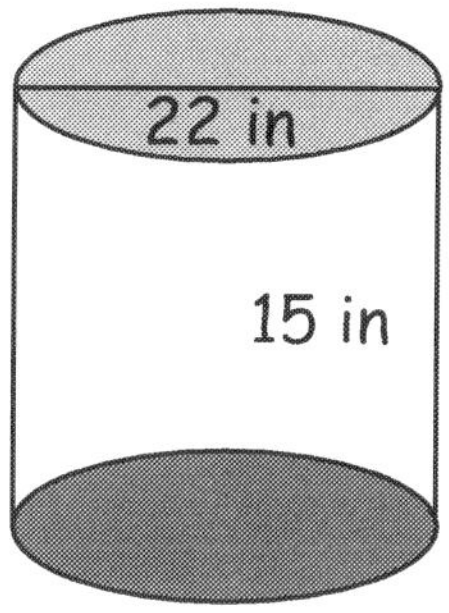

(A) 2394.9 in^3 (B) 4789.7 in^3

(C) 5702 in^3 (D) 1197.5 in^3

4. Find the volume of a cone with a diameter of 8 m and a height of 10 m.
Round the answer to the nearest tenth.

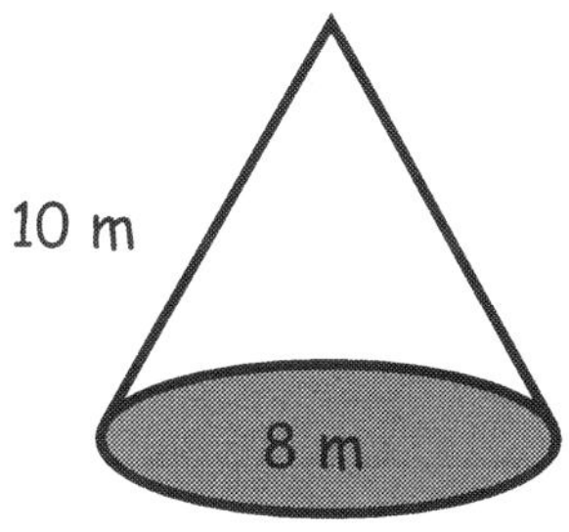

(A) 335.1 m^3 (B) 402.1 m^3

(C) 167.6 m^3 (D) 804.2 m^3

5. Find the volume of a cone with a radius of 3 km and a height of 10 km. Round the answer to the nearest tenth.

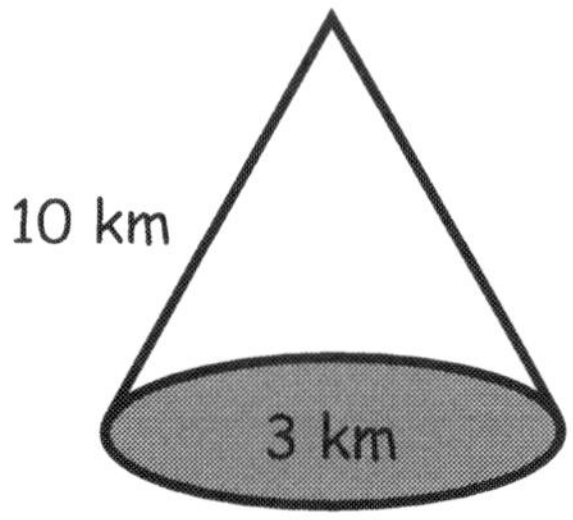

(A) 108.4 km^3 (B) 94.2 km^3

(C) 98.9 km^3 (D) 119.2 km^3

6. Find the volume of a cylinder with a radius of 14 yd and a height of 9 yd. Round the answer to the nearest tenth.

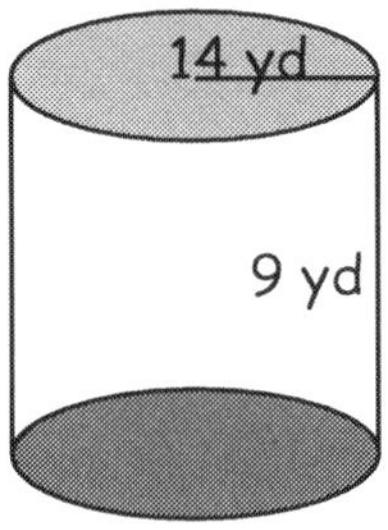

(A) 5541.8 yd^3 (B) 4876.8 yd^3

(C) 2770.9 yd^3 (D) 2438.4 yd^3

7. Find the volume of a cone with a diameter of 4 in and a height of 15 in. Round the answer to the nearest tenth.

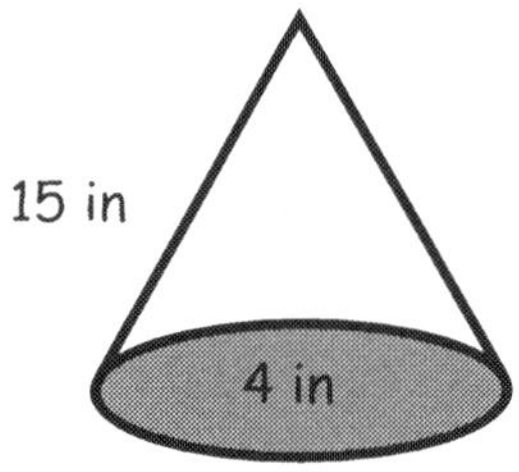

(A) 43.8 in^3 (B) 55.9 in^3

(C) 62.8 in^3 (D) 49.8 in^3

Find the value of x for the below questions.

1)

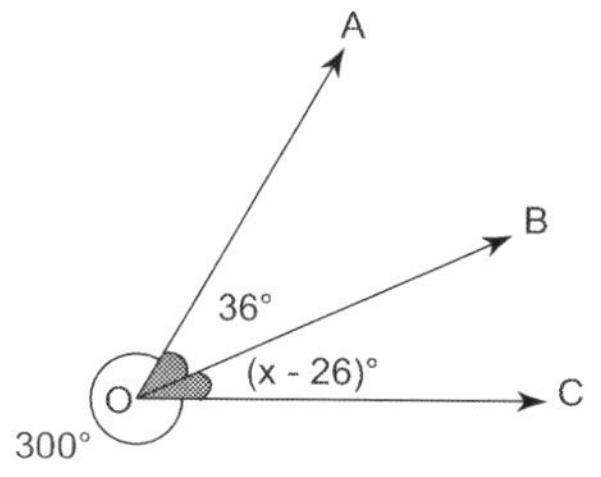

A) 48 B) 40
C) 50 D) 46

2)

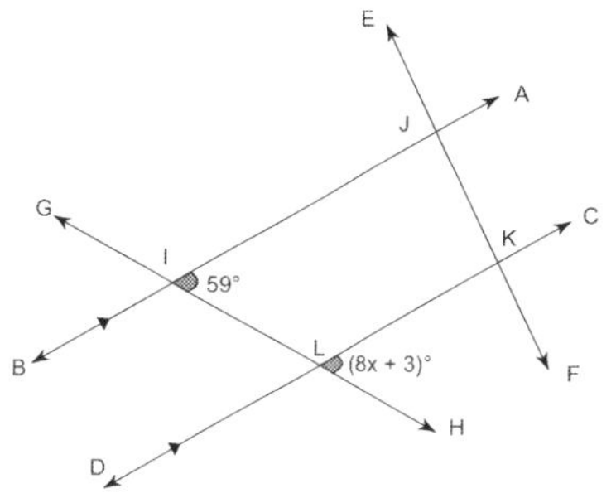

A) 16 B) 9
C) 12 D) 7

3)

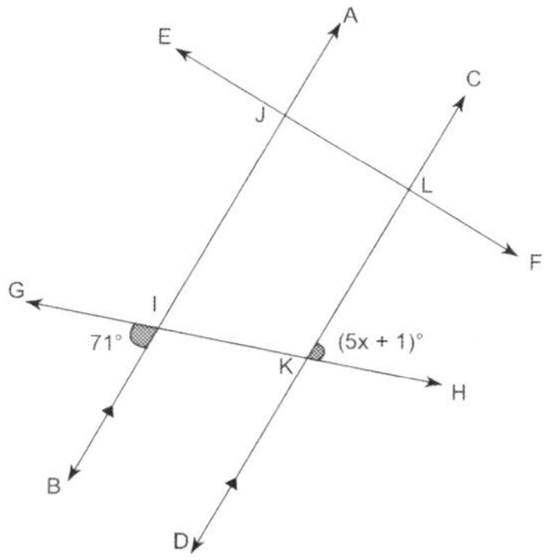

A) -1 B) 0
C) 14 D) 7

4)

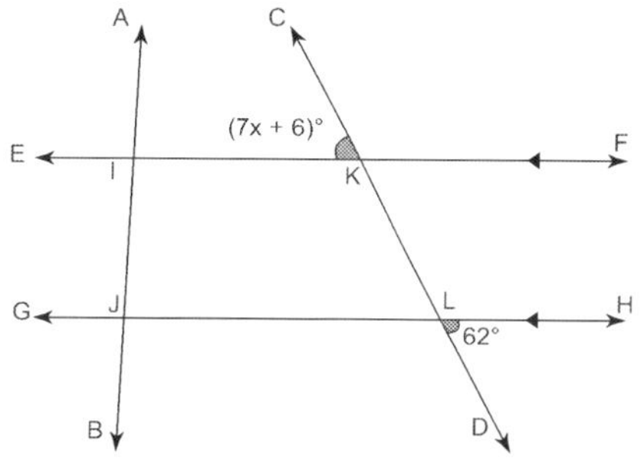

A) 14 B) 13
C) 8 D) 5

5)

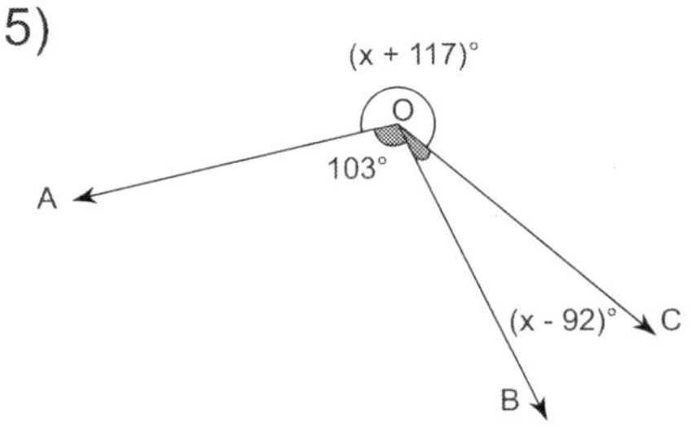

A) 122 B) 118
C) 115 D) 116

6)

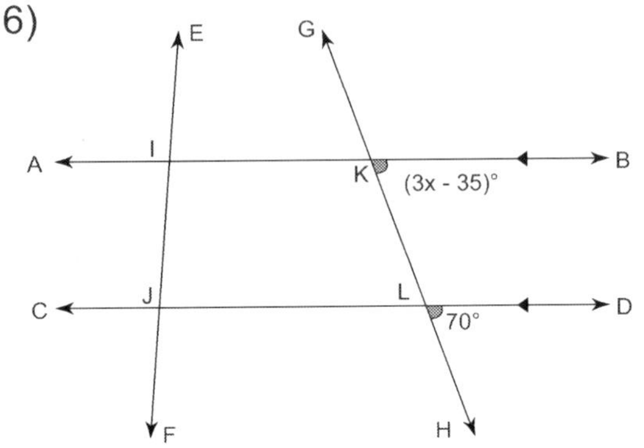

A) 35 B) 33
C) 31 D) 30

7)

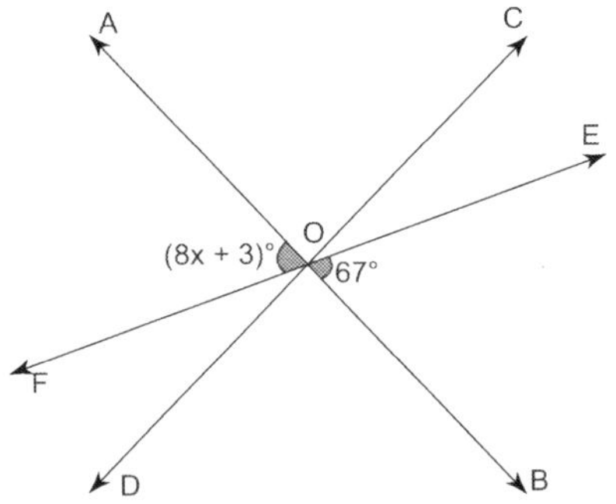

A) 7 B) 8
C) 15 D) 9

8)

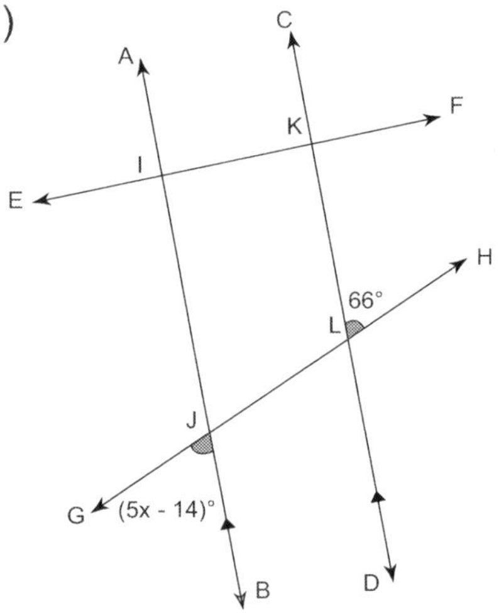

A) 16 B) 13
C) 21 D) 14

Find the measure of the ∠x for the below questions:

1)

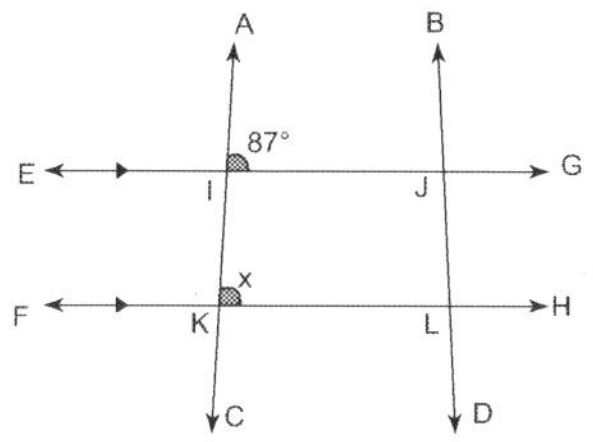

A) 155° B) 87°
C) 93° D) 3°

2)

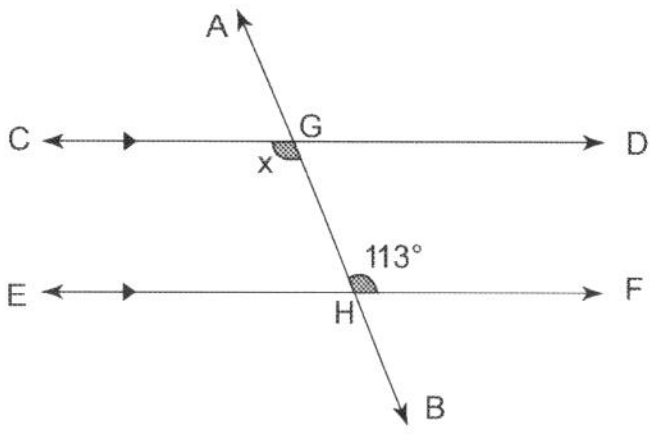

A) 130° B) 67°
C) 50° D) 113°

3)

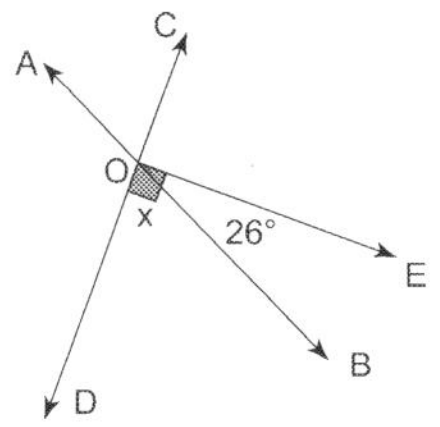

A) 93° B) 154°
C) 26° D) 64°

4)

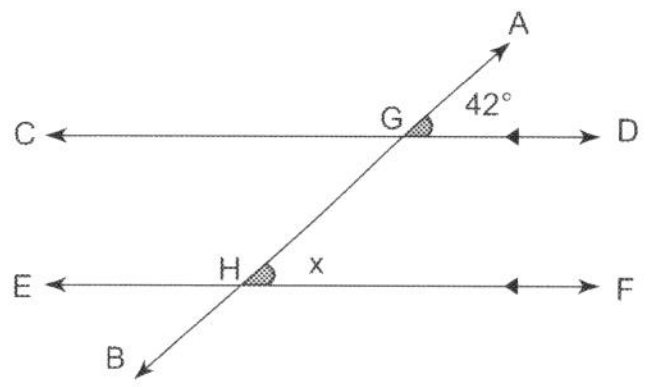

A) 42° B) 132°
C) 138° D) 48°

5)

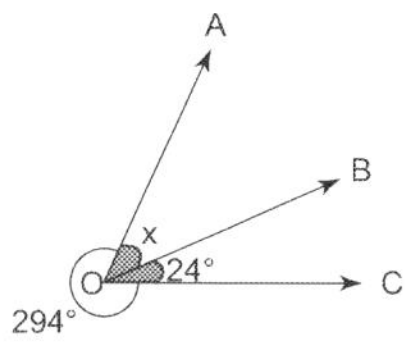

A) 42° B) 132°
C) 48° D) 138°

6)

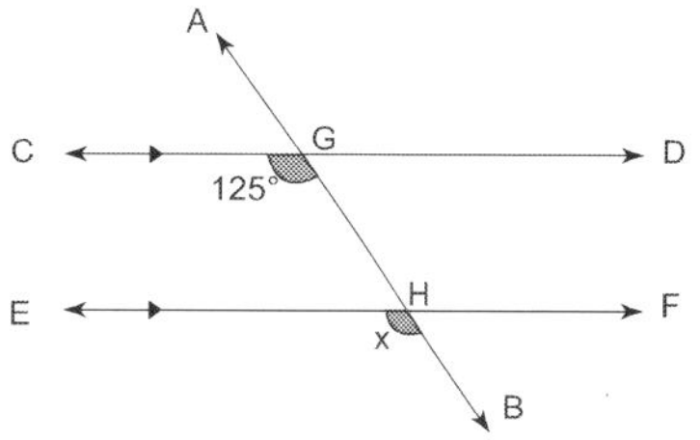

A) 145° B) 35°
C) 125° D) 55°

7)

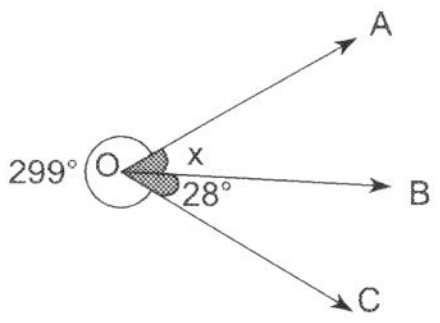

A) 57° B) 147°
C) 33° D) 123°

8)

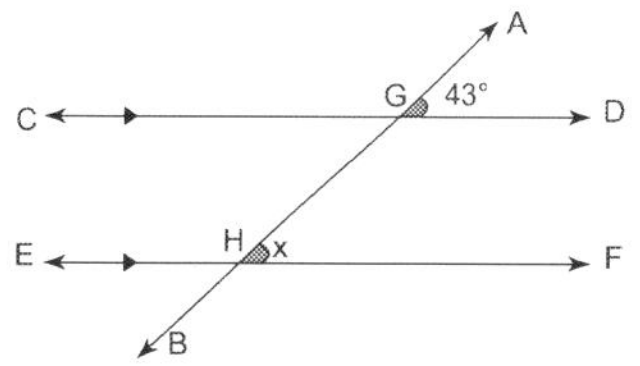

A) 146° B) 43°
C) 137° D) 56°

Describe the transformation and find the transformation rule.

1)

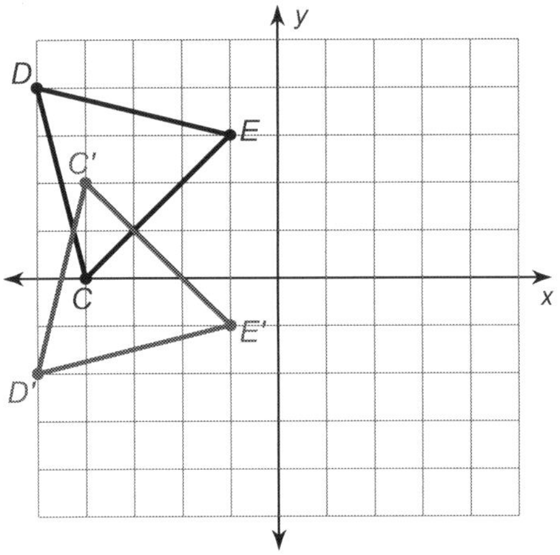

A) Reflection across $y = 1$
B) Reflection across $x = -1$
C) Reflection across $y = -1$
D) Reflection across the x-axis

2)

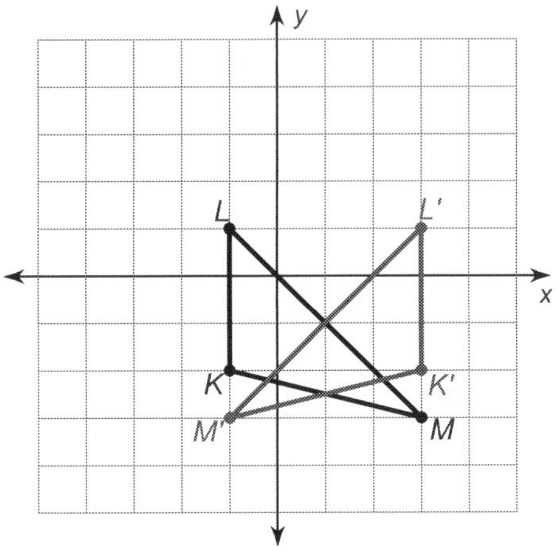

A) Reflection across $x = -2$
B) Reflection across $y = -1$
C) Reflection across $y = 1$
D) Reflection across $x = 1$

3)

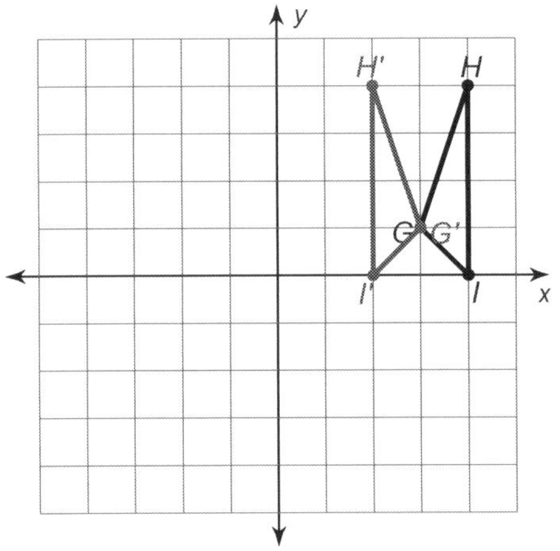

A) Reflection across x = 1
B) Reflection across $x = -3$
C) Reflection across $x = -1$
D) Reflection across $x = 3$

4)

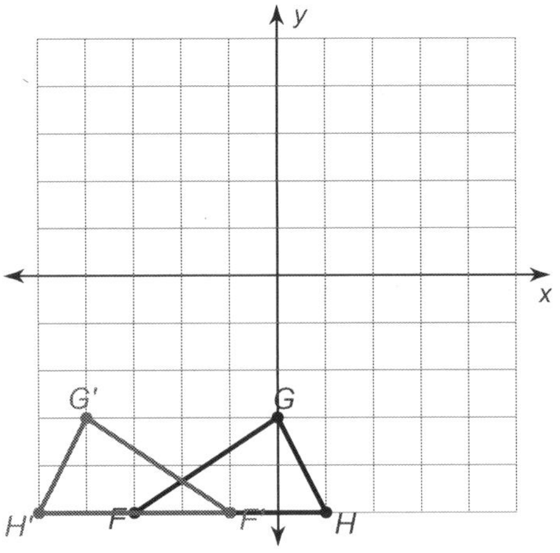

A) Rreflection across the y-axis
B) Reflection across $x = -2$
C) Reflection across $y = -2$
D) Reflection across $y = 1$

Describe the transformation and find the transformation rule.

5)

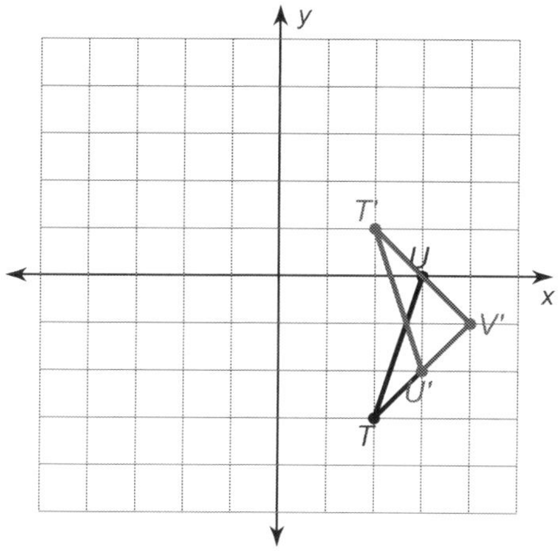

A) Reflection across $y=1$
B) Reflection across the y-axis
C) Reflection across $x=-1$
D) Reflection across $y=-1$

6)

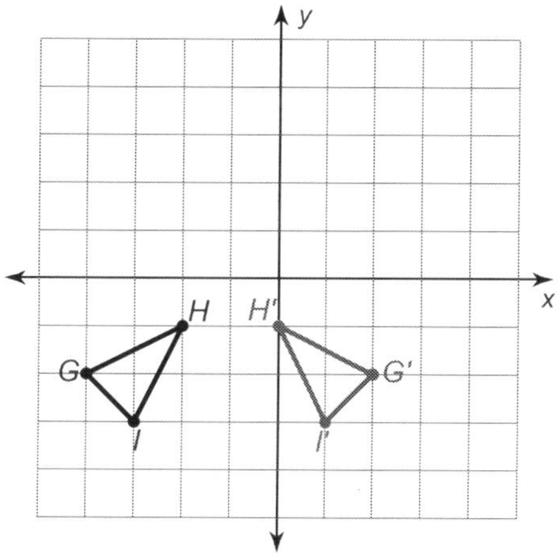

A) Reflection across $x=-1$
B) Reflection across the y-axis
C) Reflection across $x=1$
D) Reflection across $y=-1$

7)

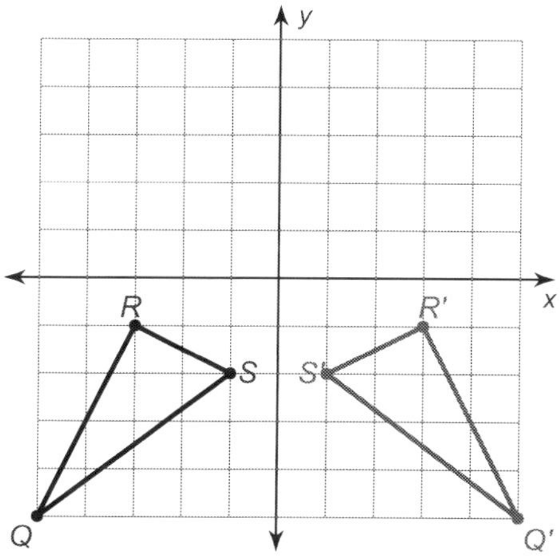

A) Reflection across $x=1$
B) Reflection across the y-axis
C) Reflection across $y=-1$
D) Reflection across $x=2$

8)

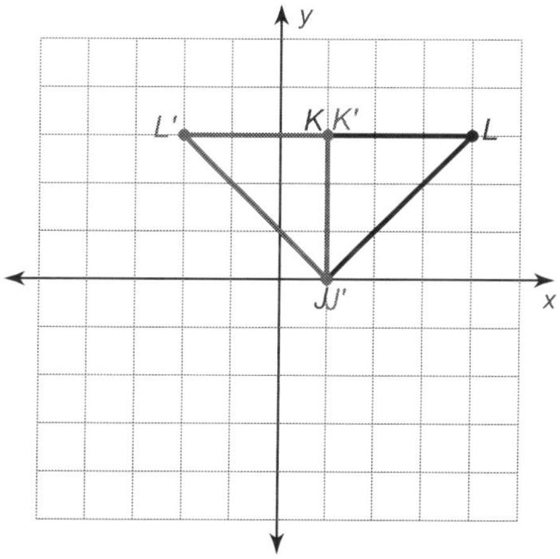

A) Reflection across the x-axis
B) Reflection across $y=2$
C) Reflection across $x=1$
D) Reflection across $x=-1$

Describe the transformation and find the transformation rule.

1)

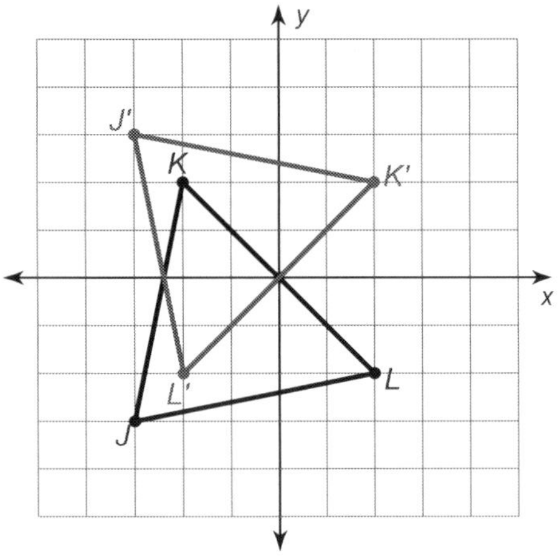

A) Rotation 180° counterclockwise about the origin
B) Translation: 1 unit right and 1 unit up
C) Rotation 90° clockwise about the origin
D) Rotation 270° about the origin

2)

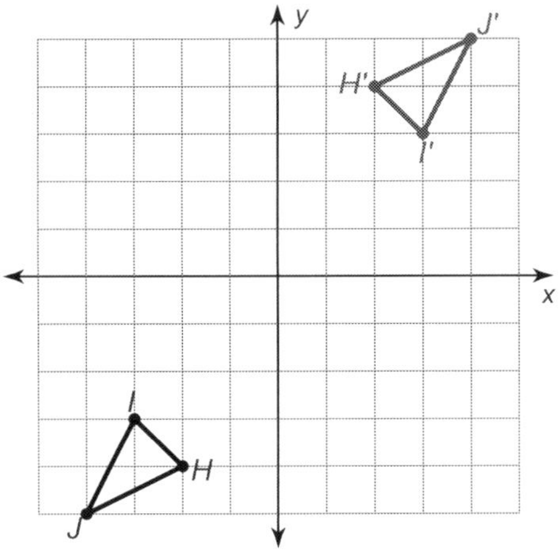

A) Rotation 180° counterclockwise about the origin
B) Rotation 180° about the origin
C) Translation: 8 units right and 5 units up
D) Rotation 90° clockwise about the origin

3)

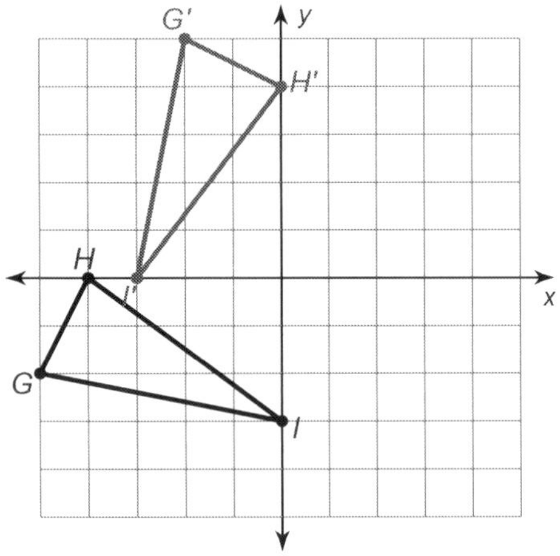

A) Rotation 90° clockwise about the origin
B) Reflection across $y=4$
C) Rotation 180° counterclockwise about the origin
D) Rotation 90° about the origin

4)

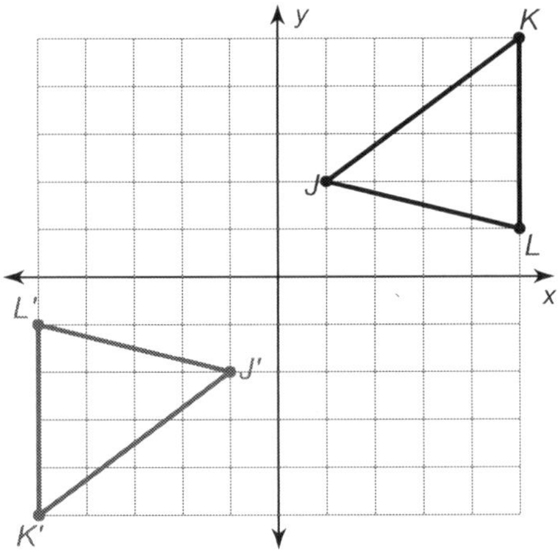

A) Rotation 90° clockwise about the origin
B) Translation: 4 units left
C) Translation: 2 units left
D) Rotation 180° about the origin

Describe the transformation and find the transformation rule.

5)

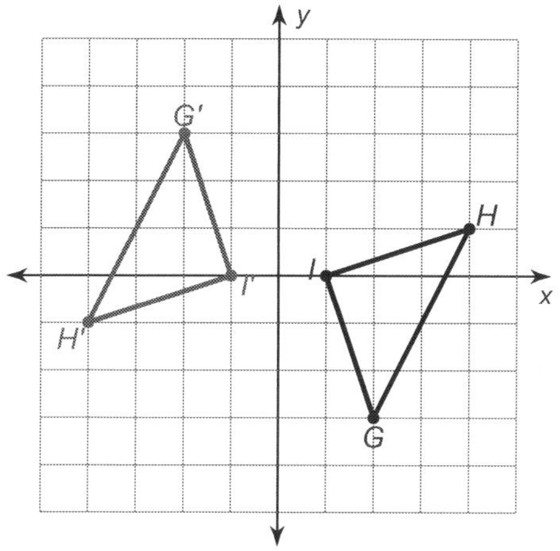

A) Rotation 90° counterclockwise
B) Rotation 180° about the origin
C) Reflection across the y-axis
D) Rotation 90° clockwise about the origin

6)

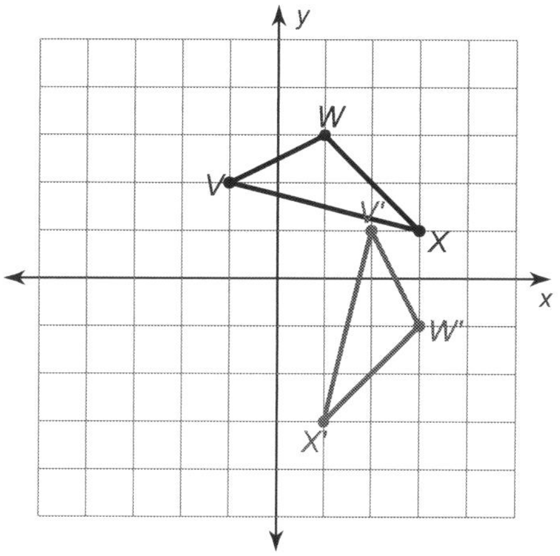

A) Rotation 180° counterclockwise about the origin
B) Translation: 3 units left and 4 units up
C) Rotation 90° clockwise about the origin
D) Rotation 180° about the origin

7)

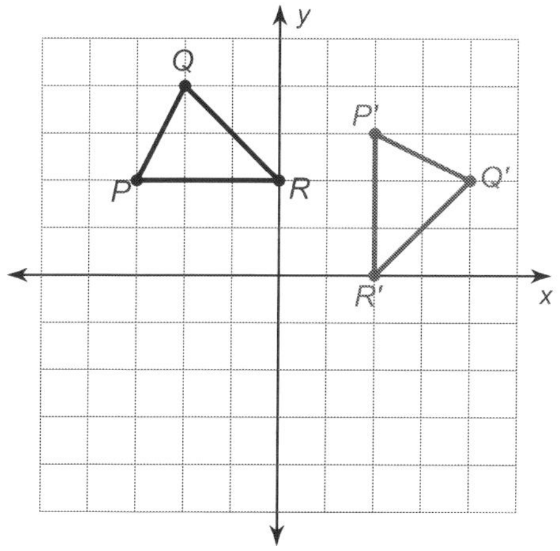

A) Rotation 270° about the origin
B) Rotation 90° clockwise about the origin
C) Rotation 180° counterclockwise about the origin
D) Reflection across $y = 5$

8)

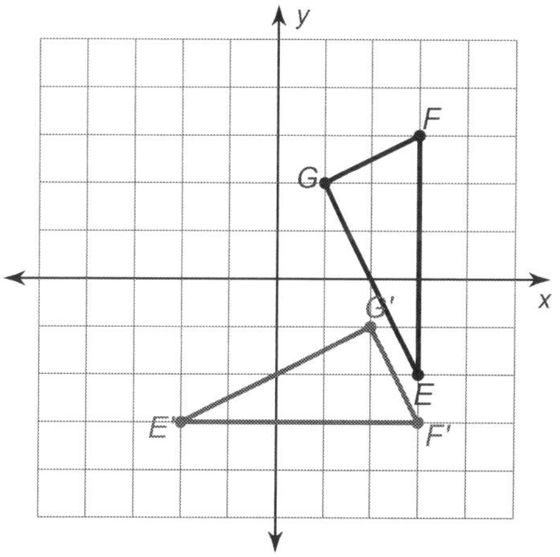

A) Rotation 180° counterclockwise about the origin
B) Translation: 5 units left and 3 units up
C) Rotation 90° clockwise about the origin
D) Rotation 270° about the origin

Write a rule to describe each transformation.

1)

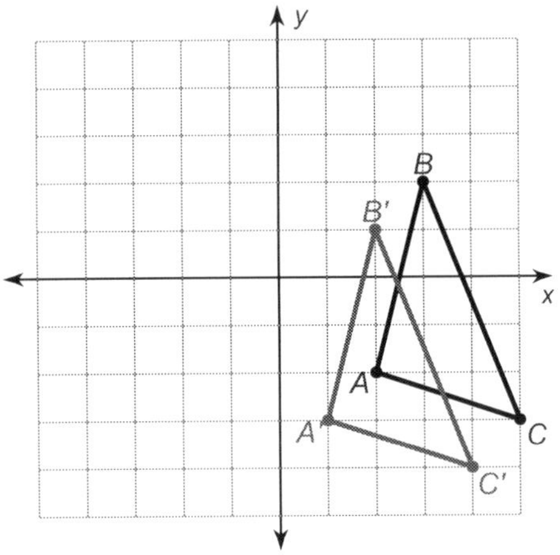

A) Translation: 4 units left and 3 units up
B) Translation: 1 unit left and 1 unit down
C) Translation: 1 unit left and 2 units up
D) Translation: 3 units left and 2 units up

2)

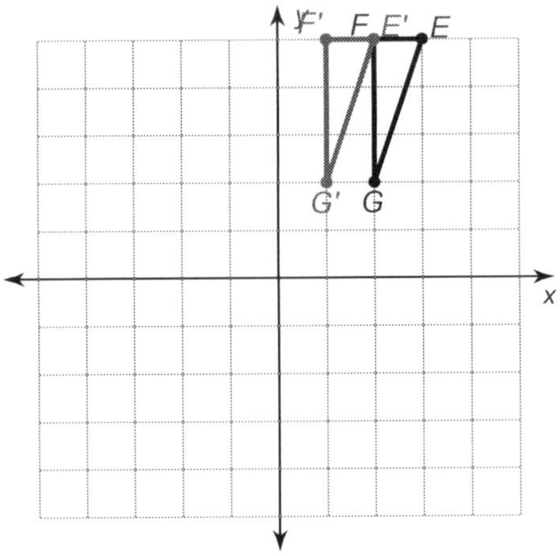

A) Translation: 2 units left and 2 units down
B) Translation: 1 unit right and 1 unit down
C) Translation: 1 unit left
D) Translation: 1 unit right and 2 units down

3)

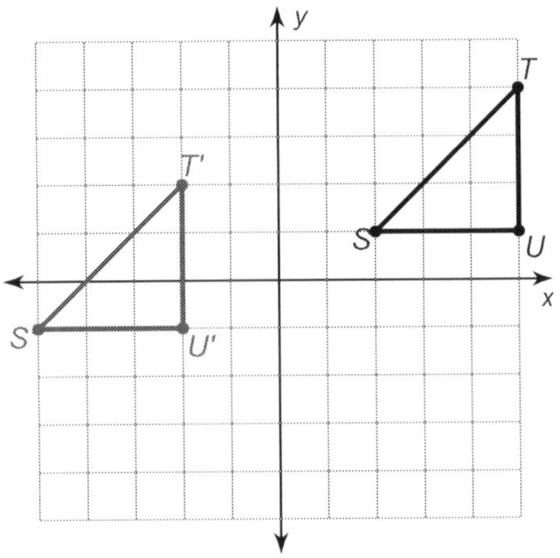

A) Translation: 7 units left
B) Translation: 7 units left and 2 units down
C) Translation: 4 units left and 2 units down
D) Translation: 4 units left and 4 units down

4)

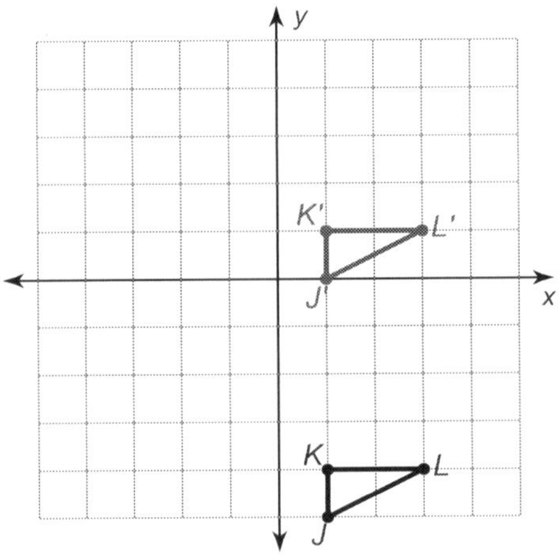

A) Translation: 5 units left and 1 unit down
B) Translation: 5 units up
C) Translation: 2 units left and 3 units up
D) Translation: 1 unit right and 6 units up

5)

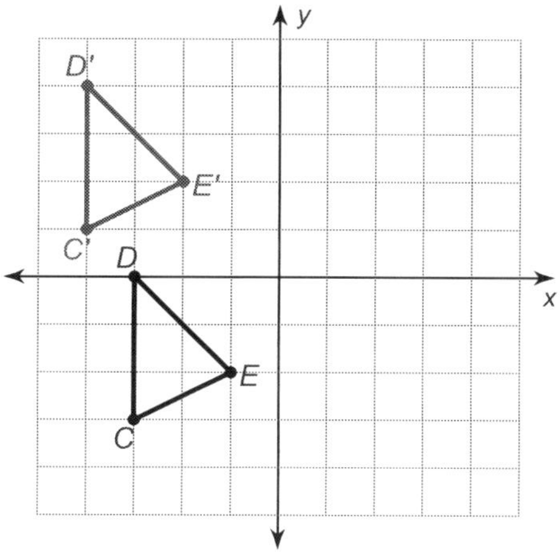

A) Translation: 1 unit left and 4 units up
B) Translation: 4 units right
C) Translation: 1 unit left and 4 units down
D) Translation: 2 units right and 4 units up

6)

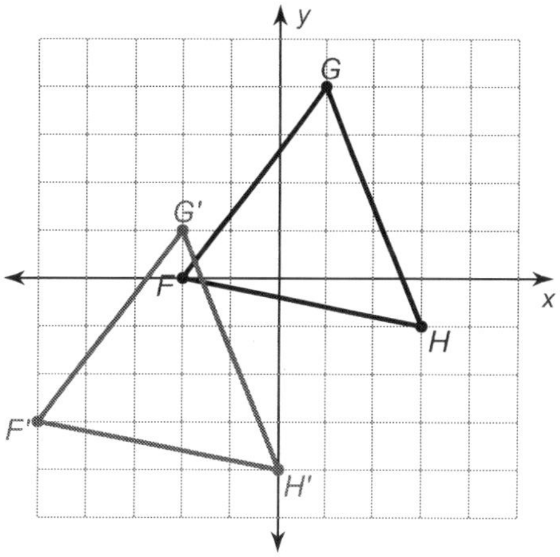

A) Translation: 3 units left and 3 units down
B) Translation: 4 units left and 3 units down
C) Translation: 3 units up
D) Translation: 3 units right

7)

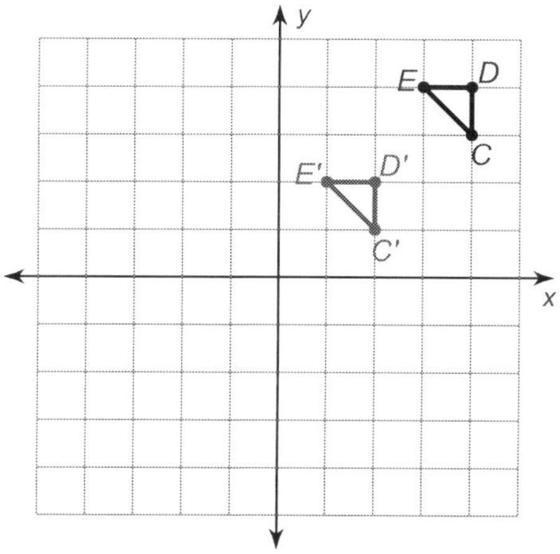

A) Translation: 5 units left and 2 units down
B) Translation: 4 units left and 2 units up
C) Translation: 2 units left and 1 unit down
D) Translation: 2 units left and 2 units down

8)

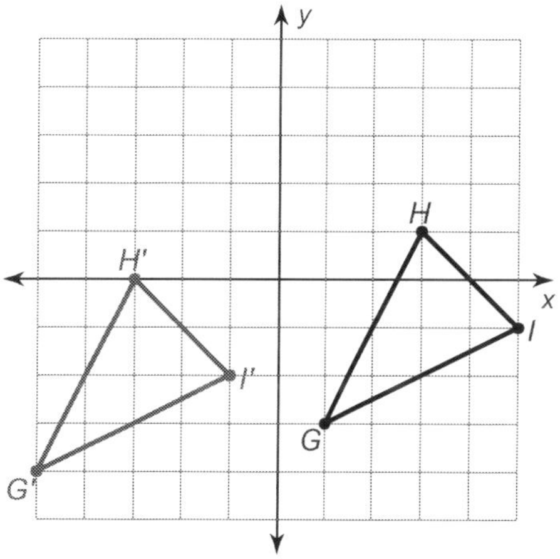

A) Translation: 6 units left and 1 unit up
B) Translation: 1 unit left and 5 units down
C) Translation: 6 units left and 1 unit down
D) Translation: 5 units right

Practice tests

SHSAT Practice Test 1

1. Evaluate $m(p + (-2)^3 + q)$; where $m = -4$, $p = -10$, and $q = 2$

(A) 69 (B) 63

(C) 64 (D) 56

2. Evaluate $|7 - 7m| = 63$, and find the sum of the solutions.

(A) 2 (B) -8

(C) 80 (D) 18

3. If $\left(\frac{1}{4}\right)^{-k} = 16^{2k+3}$, $k^2 - 2k =$

(A) 12 (B) -8

(C) 0 (D) 8

4. Find $x + y$ if $x - y = 6$ and $3x - 13y = 208$.

(A) -32 (B) -21

(C) -6 (D) 17

5 A straight line of slope $\frac{5}{2}$ is passing through the points $A(x, -2)$ and $B(1, 3)$. Find the value of x.

(A) 6 (B) -1

(C) -7 (D) 7

6. A parallelogram has a base of 24 inches and height of 15 inches. Find its area.

(A) 36 in^2 (B) 3.6 in^2

(C) 3600 in^2 (D) 360 in^2

7. If R # P = R + P - 1, then what is the value of

1 # (2 # 3) ?

(A) 8 (B) 4

(C) 11 (D) 16

8. The below box and whisker plots summerize the math test scores of grade 8 and grade 7

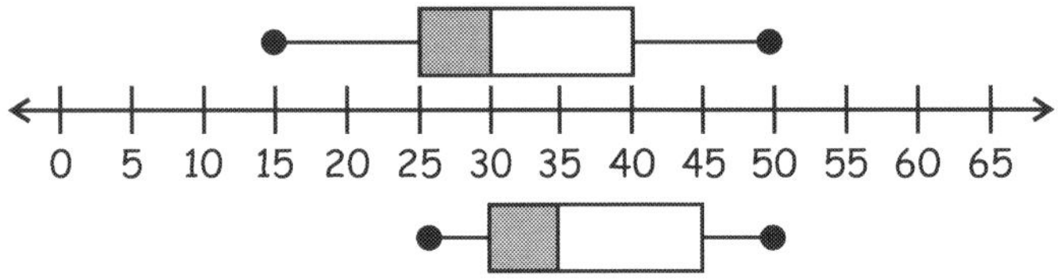

Based on the above data , which of the below statements is true ?

(A) The least score of grade 7 students is 15
(B) The highest score of grade 8 students is 45
(C) The least score of grade 8 students is 15
(D) No grade 7 students got a score of 50

9. Rik got 91, 91, 97, 94 and 95 on the math test. What is his average math score ?

(A) 93.6 (B) 93

(C) 94.5 (D) 92.6

10. Rak reads a book of 350 pages. How many 5's are there in page numbers ?

(A) 61 (B) 66

(C) 55 (D) 64

11. ZipZing company is selling their hand bags for \$175 which were originally priced as \$250 .What is the percentage of the discount offered ?

(A) 40% (B) 80%

(C) 75% (D) 30%

12. Rak puts her hand in a pile of 7 red beads, 8 green beads, and 5 gold beads. What is the probability that she chooses a red bead ?

(A) $\frac{7}{20}$ (B) $\frac{1}{4}$

(C) $\frac{8}{20}$ (D) $\frac{1}{5}$

13. If the length of the Square K is doubled, then its perimeter increases by ?

(A) 4 (B) 2

(C) 8 (D) 16

14. Pat can build a doll house in 12 hours. Lucy can build the same doll house in 16 hours. Together they can complete to build a doll house in how many hours ?

(A) 6.86 (B) 6.76

(C) 6.80 (D) 6.87

15. When you reverse digits in a certain two digit number you increase its value by 27. What is the number if the sum of its digits is 5 ?

(A) 23 (B) 19

(C) 17 (D) 14

16. In a garden, there are 10 rows and 12 columns of Apple trees. The distance between any two trees is 2 meters and a distance of one meter is left from all sides of the boundary of the garden. The length of the garden is

(A) 20 m (B) 22 m

(C) 24 m (D) 26 m

17. The compound interest for principle X is \$1025 for 2 years and the simple interest on same principle is \$1000. With the same rate of interest and number of years, What will be the principle amount for a simple interest of \$1500 ?

(A) \$2500 (B) \$3500

(C) \$3750 (D) \$4000

18. The perimeter of a rectangular field is thrice the perimeter of a square field. If one of the sides of the square field is 24 m and breadth of the rectangular field is 48 m then find the area of the rectangular field ?

(A) 4276 Sq.m (B) 5354 Sq.m

(C) 3892 Sq.m (D) 4608 Sq.m

19. Olivia bought a bag for \$36.75, which was originally priced at \$52.50. Find the discount offered by the store ?

(A) 60% (B) 30%

(C) 40% (D) 15%

20. Lydia sold a hand bag for \$32.19. She purchased the bag at \$22.99 . Find her profit percentage ?

(A) 20% (B) 9%

(C) 40% (D) 10%

21. Claire is purchasing a land priced at \$16,500 and she pays a tax of 5%. Find the selling price of the land.

(A) \$17,325.00 (B) \$14,025.00

(C) \$13,200.00 (D) \$18,150.00

22. Dan added 18 instead of 13.6 in his calculation. Find the percentage change in his calculation

(A) 24.4% increase (B) 4.4% decrease

(C) 132.4% increase (D) 32.4% increase

23. A farmhouse shelters 14 animals. Some are horses and some are geese. Altogether there are 48 legs. How many geese are there in the farmhouse ?

(A) 12 geese (B) 4 geese

(C) 10 geese (D) 11 geese

24. Sam took a loan of \$53,300 at 6% interest compounded semi annually for 8 years. Find the amount he needs to pay at the end of the loan term.

(A) \$85,530.85 (B) \$135,400.74

(C) \$84,952.10 (D) \$78,884.00

25. Find the missing terms in the below arithmetic sequences.

..., -10, ___, ___, ___, 14, ...

(A) -9, -14, -19 (B) 2, 8, 14

(C) -10, -16, -22 (D) -4, 2, 8

26. Find the missing terms in the below arithmetic sequences.

..., 8, ___, ___, ___, ___, ___, 1208, ...

(A) -192, -392, -592, -792, -992

(B) 208, 408, 608, 808, 1008

(C) -392, -592, -792, -992, -1192,

(D) -592, -792, -992, -1192, -1392

27. Evaluate $\dfrac{21+3}{(-4)-(-10)}$

(A) 8

(B) 1

(C) -4

(D) 4

28. Solve the proportion $\dfrac{x}{3}=\dfrac{x+2}{6}$

(A) 106

(B) 2

(C) 8.6

(D) -1.9

29. How many lb of brand X coffee which costs $25/lb must be added to 14 lb of brand Y coffee which costs $10/lb to make Rak's special coffee blend which costs $11/lb?

(A) 3 lbs

(B) 4 lbs

(C) 6 lbs

(D) 1 lb

30. Shan went to drop his daughter at the train station and came back home. The trip there took five hours and the trip back took four hours. What was Shan's average speed on the trip if he averaged 60 km/h on the return trip?

(A) 30 km/h

(B) 20 km/h

(B) 50 km/h

(D) 48 km/h

31. If $P(B)=\dfrac{13}{20}$ $P(A \text{ and } B)=\dfrac{143}{400}$ *then* $P(A)=$?

(A) $\dfrac{11}{80}$

(B) $\dfrac{11}{20}$

(C) $\dfrac{63}{400}$

(D) $\dfrac{1}{20}$

32. Henry flipped a coin three times and then rolls a fair six-sided dice five times. Find the probability of the coin showing heads-up every time and the dice showing an even number every time.

(A) $\frac{28}{121}$ (B) $\frac{1}{256}$

(C) $\frac{8}{165}$ (D) $\frac{10}{273}$

33. A magic bag has eight nickels and seven dimes. Rak picks a coin from the bag and gives it her brother Rik. She picks another coin and gives it to him as well. Find the probability that the first coin is a nickel and the second coin is a dime.

(A) $\frac{1}{36}$ (B) $\frac{1}{8}$

(C) $\frac{1}{27}$ (D) $\frac{4}{15}$

34. Working together, Rik and Rak can harvest a field in 8.06 hours. Had she done it alone it would have taken Rak 19 hours. How long would it take Rik to do it alone ?

(A) 13.69 hours (B) 13.99 hours

(C) 15 hours (D) 16.02 hours

35. Two pipes P and Q can fill the tank within 12 hours and 18 hours respectively. Pipe R can empty the tank in 36 hrs. If all three pipes are opened, find the time taken to fill the tank?

(A) 8 hours (B) 9 hours

(C) 10hours (D) 12 hours

36. A cube is painted with red, black and green colors. Opposite faces are painted with same color. How many small cubes are there where one face is green and other one is either black or red ?

(A) 28 (B) 8

(C) 16 (D) 24

37. For the given series below

-4 , -2 , 0 , . . .

If $\begin{cases} a(1) = A \\ a(n) = a(n-1) + B \\ A = ?, B = ? \end{cases}$, find $A^2 + B^2$

(A) 20 (B) 2

(C) 4 (D) 36

38. Danny has a 6 sided dice. marked by number four on four sides and number two is marked on the remaining two sides. Which statement best describes Danny's chance of winning the game by getting the number 4?

(A) Certainly will win

(B) Certainly will lose

(C) Most likely will win

(D) Most likely will lose

39. The formula for blood flow rate is given by $F = \dfrac{P_1 - P_2}{r}$

where F is the flow rate, P_1 the initial pressure , P2 the final pressure and r the resistance created by blood vessel size. Which formula can not be derived from the given formula ?

(A) $P_1 = Fr + P_2$

(B) $P_2 = P_1 - Fr$

(C) $r = F(P_2 - P_1)$

(D) $r = \dfrac{P_1 - P_2}{F}$

40. A box contains 10 red beads , 12 gold beads , 15 black beads , 13 green beads. Chuck picks bead from the box , what is the probability that it will be a gold or a black bead ?

(A) 54 %

(B) 43 %

(C) 59 %

(D) 68 %

41. The parallel side of a trapezium are 29 inches and 11 inches and non parallel side of length 13 inches and 15 inches. Find the distance between the parallel sides.

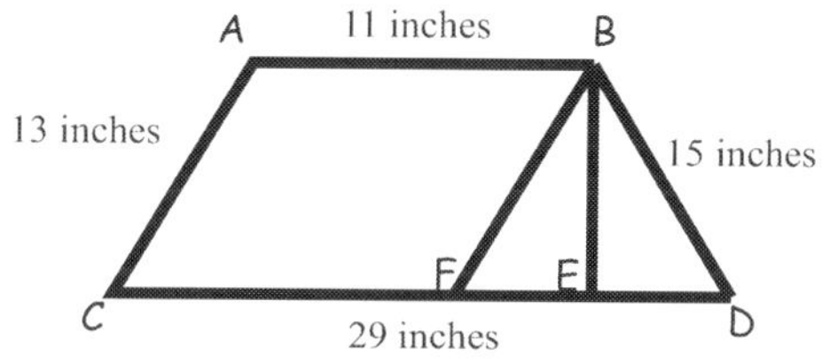

(A) 15 inches

(B) 11 inches

(C) 12 inches

(D) 14 inches

42. In a string of beads for every 2 gold beads there are 3 green beads. What is the color of 223rd bead ?

(A) Gold

(B) Green

(C) Blue

(D) None

43. Sam took a loan of $205 at 12% interest for 7 years. Find the amount he needs to pay at the end of the loan term.

(A) $377.20

(B) $453.19

(C) $248.19

(D) $229.60

44. Simplify the expression 7(m + 4) + 3(5m + 3)

(A) 22m + 37
(B) 28 - m
(C) 70 - 110m
(D) 22m + 27

45. Simplify the expression 8 + 19(1 + 5r)

(A) -201r + 216
(B) 27 + 95r
(C) 27 + 75r
(D) -67 + 60r

46. Simplify the expression and plot the graph $\frac{-12+n}{2} < -17$

(A) n < -41 : 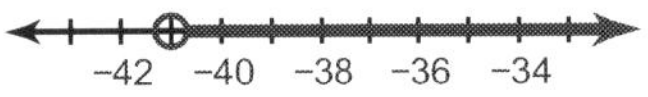

(B) n < -22 : 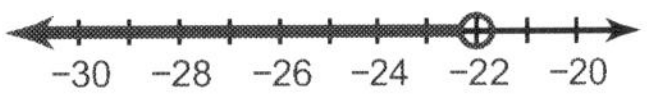

(C) n < -41 : 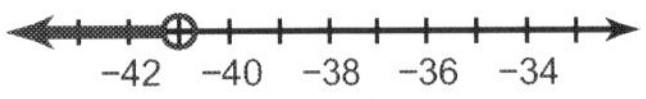

(D) n < -17 : 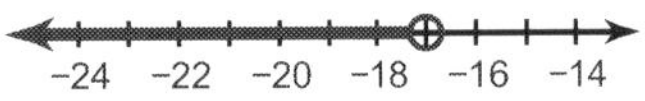

47. Simplify the inequality and graph the solution

-2 - 12k < 106 and 5 - 4k ≥ 29

(A) -9 < k ≤ -6 : 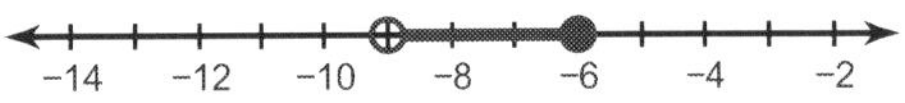

(B) k ≥ -6 : 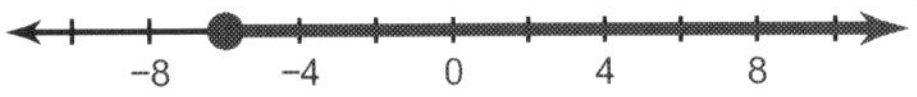

(C) k < 9 : 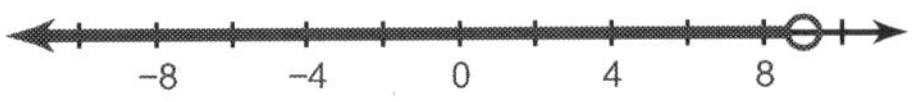

(D) k ≤ -6 : 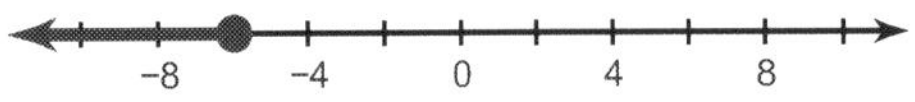

48. Which of the below represents all possible unique factors of 58

(A) 1,2,58
(B) 1,16,29,58
(C) 1,2,29,58
(D) 1,29,58

49. Find the value of x from the figure given below

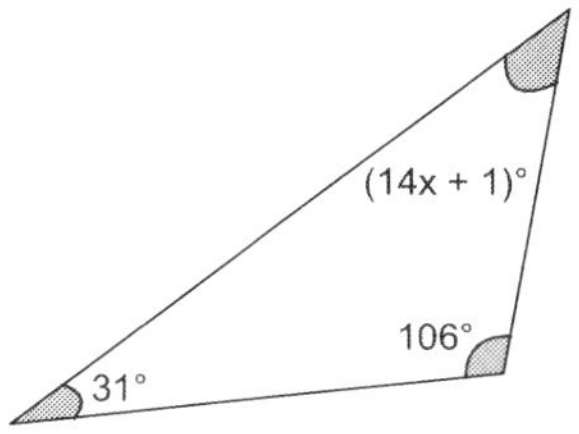

(A) 1
(B) 8
(C) 3
(D) 9

50. Find the value of x from the figure given below

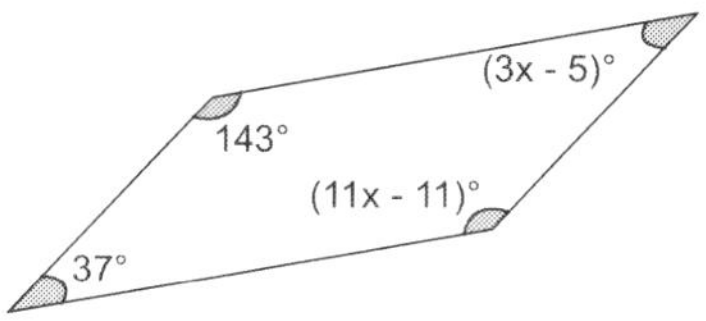

(A) 8
(B) 9
(C) 14
(D) 16

51. Find the value of x from the figure given below

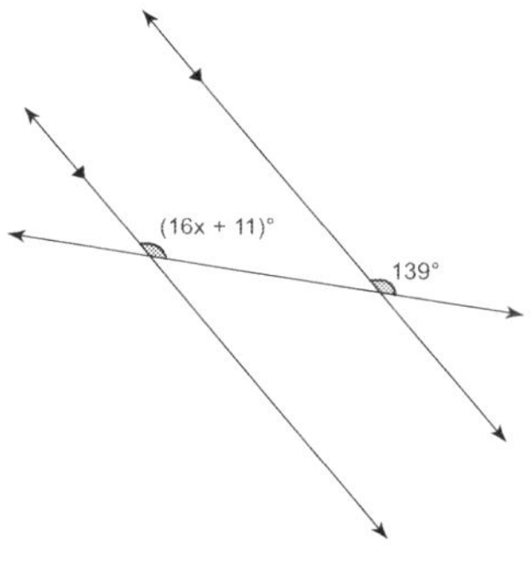

(A) 8
(B) 10
(C) 13
(D) -5

52. Solve the inequality and graph its solution $\frac{|-2n-2|}{9} > 4$

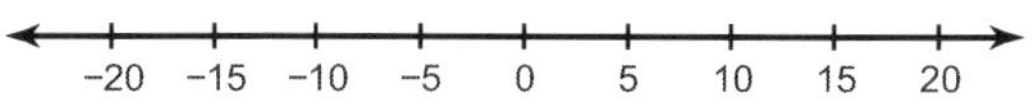

53. Find the midpoint of the line segment $\overline{AB}$ given below

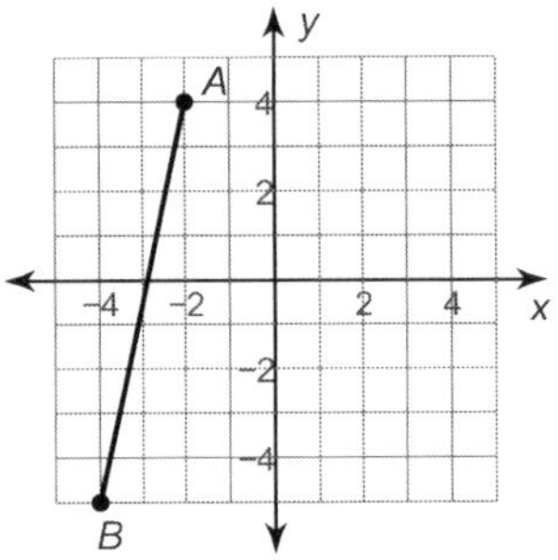

(A) (1,4.5)   (B) (-3,-0.5)

(C) (1,-4.5)   (D) (-6,-14)

54. James reduced the size of a triangle to a height of 1.1 in. What is the new width if it was originally 6.6 in tall and 19.2 in wide?

(A) 3.2 in   (B) 0.2 in

(C) 6.6 in   (D) 2.8 in

55. With the help of scientists, farmers in the United States have been able to produce more and more grain per hectare each year. Here are the crop yields for several years:

Year	Yield (kg/hectare)
1964	2,760
1969	3,280
1976	3,620
1990	4,820
2002	5,710
2007	6,900

The crop yield can be described by the equation $y = 88.2x - 171000$ where x is the year and y is the grain yield in kilograms per hectare (kg/ha).

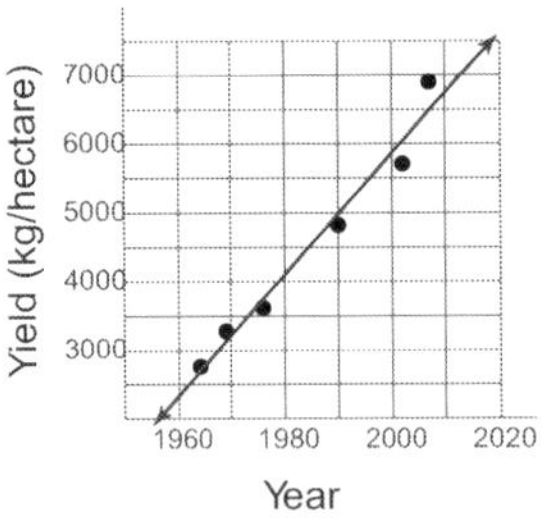

The model indicates that a crop yield of 4800 kg/hectare was achieved in what year? Round your answer to the nearest year.

(A) 1993   (B) 2162

(C) 1847   (D) 1906

56. Evaluate the below expression

$$-1 - \left(-\frac{1}{2}\right) - \left(-2\frac{1}{4}\right)$$

BR High school conducted a survey on 1800 students for their favorite hobbies. The below pie chart represents percentage of student votes. Answer the questions 57 and 58.

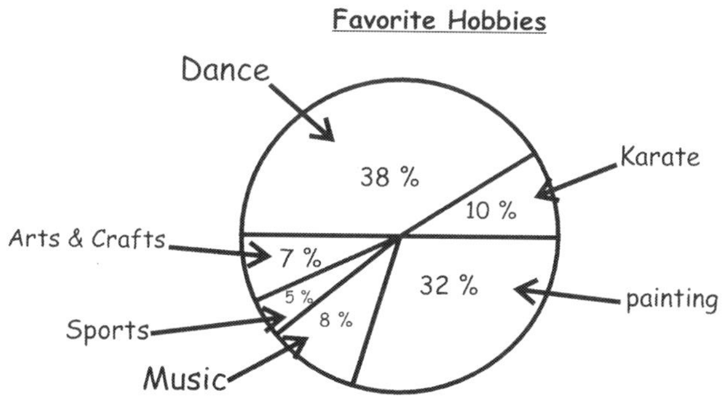

57. Find the number of students who voted for dancing

(A) 648 (B) 684

(C) 864 (D) 664

58. Find the ratio between the number of student votes for painting and karate.

(A) 3:7 (B) 5:16

(C) 16:5 (D) 5:7

SHSAT Practice Test 2

1. Evaluate $p(1 - m) - \frac{m}{4}$; where m = 4, and p = 7

(A) −22 (B) −31

(C) −17 (D) −27

2. Find the sum of the solutions of $|x - 6| = 5$

(A) 6 (B) $\frac{9}{5}$

(C) $\frac{-5}{3}$ (D) 12

3. If $3^{-2x} . 3^{x+1} = \frac{1}{81}$, find the value of $x^2 + 1$

(A) $-\frac{1}{8}$ (B) 26

(C) 6 (D) $\frac{25}{9}$

4. Find $x^2 + y^2$ if $x + y = -6$ and $11x - y = 18$

(A) 50 (B) 17

(C) 53 (D) 28

5. A straight line of slope $-\frac{1}{2}$ is passing through the points $A(-9, -2)$ and $B(x, -9)$. Find the value of x.

(A) -2 (B) 1

(C) 5 (D) -5

6. Simplify $a - (a^2 + bc)$; where a = 3 , b = -15 and c = 5

(A) 61 (B) 78

(C) 58 (D) 69

7. Hazel purchased a book at $6.95 and sells it for 35% price increase. Find the selling price of the book ?

(A) $4.52 (B) $9.38

(C) $2.43 (D) $6.60

8. Working alone it takes Beth 15 hours to harvest a field. Mei can harvest the same field in 9 hours. Find how long it would take them if they worked together ?

(A) 4.3 hours (B) 7.15 hours

(C) 5.63 hours (D) 5.82 hours

9. Jason is preparing to bake a new year cake. Her recipe needs $5\frac{1}{9}$ cups of sugar. Instead she adds $5\frac{5}{6}$ cups of sugar. How many extra cups of sugar did she add ?

(A) $5\frac{1}{9}$ (B) $5\frac{1}{6}$

(C) $10\frac{17}{18}$ (D) $\frac{13}{18}$

10. Find the volume of a cylinder with a radius of 10 cm and a height of 16 cm. Answer to the nearest tenth

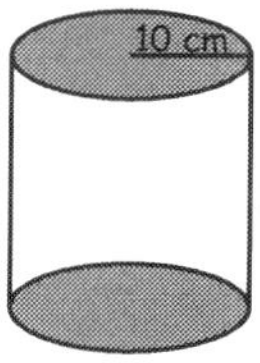

(A) 3096.7 cm^3 (B) 5026.5 cm^3

(C) 2789.8 cm^3 (D) 2513.3 cm^3

11. Which of the below options form a right angle triangle ?

(A) 28 cm , 45 cm , 46 cm (B) 39 cm , 80 cm , 81 cm

(C) 32 cm , 60 cm , 68 cm (D) 30 cm , 40 cm , 41 cm

12. Fix Nut company charges $25.65 for fixing the air in car tyres, which was originally priced at $28.50. What is the percentage of the discount ?

(A) 30% (B) 20%

(C) 15% (D) 10%

13. If we toss a coin 4 times, how many different outcomes are possible ?

(A) 8 (B) 10

(C) 16 (D) 12

14. Which of the below options describes the transformation of the below figure GEF ?

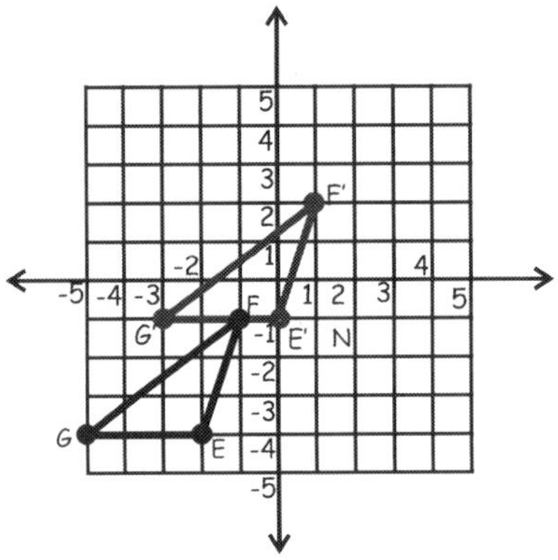

(A) reflection across y = -1

(B) translation of 2 units right and 3 units up

(C) rotation of 90^0 clockwise about the origin

(D) translation of 5 units right and 1 unit down

15. The below box and whisker plots summerize the science quiz scores of grade 8 and grade 7

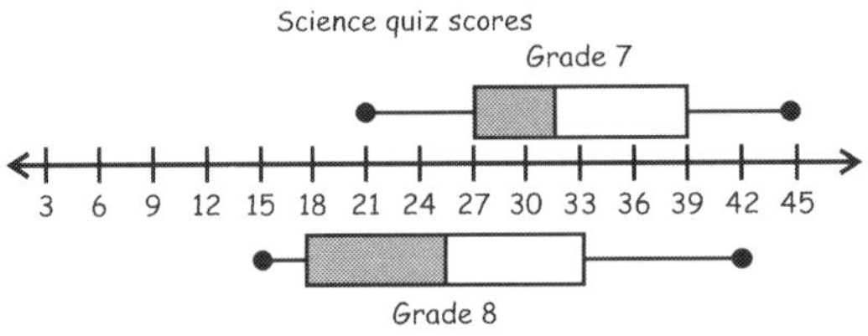

Based on the above data , which of the below statements is true ?

(A) The least score of grade 8 students is 21
(B) The highest score of grade 8 students is 55
(C) The highest score of grade 7 students is 45
(D) The least score of grade 7 students is 15

16. Maria, Bina and Cathy entered into a partnership with investment of \$ 24000, \$ 32000 and \$ 48000 respectively. After one year Maria withdrew \$4000 and Bina invested \$8000 more. And after one more year Maria invested \$15000 more and Cathy withdrew \$8000. At the end of three years, they earned a profit of \$114450. Find the share of Maria in the total profit ?

(A) \$33500 (B) \$30250

(C) \$27650 (D) \$35750

17. Zack rows from shore A to B which is at a distance of 120 km and come back in 8 hours. He can row 20 km with the stream in the same time as 12 km against the stream. The rate of the stream is?

(A) 12 km/hr (B) 16 km/hr

(C) 10 km/hr (D) 8 km/hr

18. Jackson sells two types of candles. He sells small candles for \$12 and big candles for \$14. He can only sell a maximum of 40 candles per week. "m" represents the number of small candles and "g" represents the number of big candles. If he needs to earn at least \$250 this week , which system of inequalities represents the given conditions ?

(A) m + g <= 40 / 12m + 14g >= 250
(B) m + g >= 250 / 12m + 14g <= 40
(C) m + g <= 250 / 12m + 14g >= 40
(D) m + g <= 40 / 12m + 14g <= 250

19. A box contains 8 puzzles , 4 dictionaries , 3 story books , 10 magazines Rosy picks an item from the box , what is the probability that it will be a puzzle or a story book ?

(A) 30 % (B) 39 %

(C) 22 % (D) 44 %

20. Find the approximate length of the missing side AC from the below figure

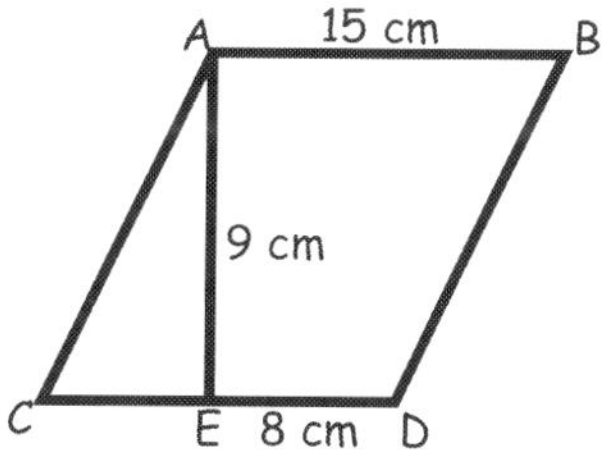

(A) 12 cm (B) 15 cm

(C) 11 cm (D) 21 cm

21. The parallel side of a trapezium are 35 inches and 19 inches and non parallel side of length 17 inches and 17 inches. Find the distance between the parallel sides.

(A) 15 inches (B) 27 inches

(C) 24 inches (D) 13 inches

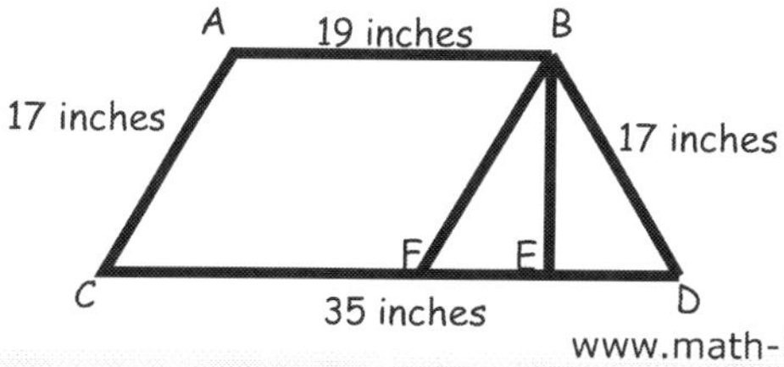

22. In a class, 20% of the members own only two cars each, 40% of the remaining own three cars each and the remaining members own only one car each. Find the percentage of members having one car?

(A) 52% (B) 48%

(C) 60% (D) 80%

23. In an election between two contestants, 20 % of the voters didn't vote and 160 votes are declared invalid. The winner gets 39 % of the votes and declared win by 310 votes. Then find the number of voters in the voting list?

(A) 6000 (B) 4500

(C) 6250 (D) 6500

24. Robert purchased a X - Box at $629.99 and sells it at a 55% price increase. Find the selling price of the X - Box.

(A) $566.99 (B) $346.49

(C) $283.50 (D) $976.48

25. Find the sum of the series as given below

$$a_1 = 8 , a_n = 36 , n = 15$$

(A) 337 (B) 333

(C) 660 (D) 330

26. Rak added number 4 instead of 95 in her calculation. Find the percentage change in her calculations.

(A) 94.8% increase (B) 95.8% decrease

(C) 132.4% increase (D) 32.4% increase

27. Beth took a loan of $460 at 10% interest for 9 years. Find the amount she needs to pay at the end of the loan term

(A) $506.00 (B) $1,084.66

(C) $414.00 (D) $874.00

28. Jack took a loan of $26,000 at 6% interest compounded semi annually for 8 years. Find the amount he needs to pay at the end of the loan term.

(A) $35,692.43 (B) $30,463.14

(C) $30,487.04 (D) $30,160.00

29. A farmhouse shelters 14 animals. Some are horses and some are geese. Altogether there are 48 legs. How many horses are there in the farmhouse ?

(A) 4 horses (B) 10 horses

(C) 2 horses (D) 3 horses

30. Find the missing terms in the below arithmetic sequences

..., 36, ___, ___, 636, ...

(A) 236, 436 (B) 239, 439

(C) 238, 438 (D) 439, 639

31. Find the missing terms in the below sequences

..., -5, ___, ___, ___, ___, ___, 49, ...

(A) 4, 13, 22, 31, 40 (B) 3, 11, 19, 27, 35

(C) 12, 21, 30, 39, 48 (D) 3, 12, 21, 30, 39

32. Evaluate $-\frac{1}{(-3)-2\times(-1)}$

(A) 1 (B) 2

(C) -4 (D) 9

33. Solve the proportion $\frac{7}{14}=\frac{r}{r+9}$

(A) -1 (B) 9

(C) 5.94 (C) 8.6

34. How many lbs of almonds which cost \$6/lb must be added to 3 lbs of cashews which cost \$11/lbs to make mixed nuts which cost \$9/lb?

(A) 3 lbs (B) 5 lbs

(C) 6 lbs (D) 2 lbs

36. A private jet made a round trip from Virginia to New york. It took eight hours to reach and the trip back took ten hours. If the jet averaged 46 km/h faster to reach than on the return trip. Find the private jet's average speed on the outbound trip.

(A) 335 km/h (B) 320 km/h

(C) 110 km/h (D) 230 km/h

37. If $P(B) = \frac{3}{5}$ $P(A|B) = \frac{3}{8}$ then $P(A \text{ and } B) = ?$

A) $\frac{143}{400}$ (B) $\frac{9}{40}$

C) $\frac{11}{20}$ (D) $\frac{3}{20}$

38. Rita flips a coin five times and then rolls a fair six-sided dice once. Find the probability of the coin landing heads-up every time and the dice showing a four

(A) $\frac{1}{27}$ (B) $\frac{1}{1024}$

(C) $\frac{28}{135}$ (D) $\frac{1}{192}$

39. There are five boys and five girls in a class. The teacher randomly selects three different students to answer questions. Find the probability of choosing the first student a boy, the second student is a girl, and the third student a girl.

(A) $\frac{5}{36}$ (B) $\frac{1}{28}$

(C) $\frac{10}{39}$ (D) $\frac{1}{4}$

40. 5 years ago Rita and Geetha's ages are in the ratio of 5:6 and 5 years hence the ratio of their ages becomes 6:7. Find the current age of Geetha ?

(A) 60 (B) 65

(C) 70 (D) 55

41. Find the volume of the cylinder whose radius is 2 times of breadth of rectangle whose perimeter is 90 cm, and length and breadth is in the ratio of 2:1 and cylinder height is equal side of square whose diagonal value is $20\sqrt{2}$ cm ?

(A) 47852 cm^3 (B) 26871 cm^3

(C) 56571 cm^3 (D) 69879 cm^3

42. *Find* $f(3)$ *if* $f(x) = f(x-1) + 2.x^2$ *and we know that* $f(1) = -2$

(A) 6 (B) 24

(C) 30 (D) 18

43. Raksha's average score on the two Social tests is 94. If she wants to get an average score of 95 in three tests of this quarter, what should be her score in the third test ?

(A) 89 (B) 95

(C) 90 (D) 97

44. Simplify the expression (-k + 9) - 2(7k + 6)

(A) -15k - 21 (B) -76 + 67k

(C) -14k - 21 (D) -7k - 21

45. Simplify the expression -18 - 6(-10n + 12)

(A) -21n - 140 (B) -32n - 140

(C) -90 + 60n (D) -21n - 151

46. Simplify the expression and plot the graph 98 -1 + 9x

(A) $x \le 11$:

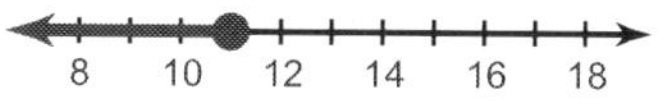

(B) $x \le -28$:

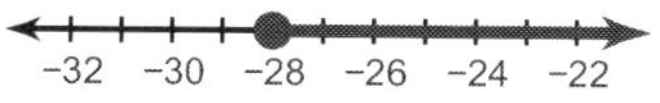

(C) $x \le -28$:

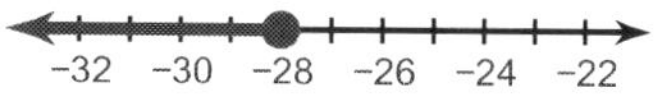

(D) $x \le -28$:

47. Simplify the inequality and graph the solution -2m - 3 ≤ -27 or -m + 5 > 17

(A) m ≥ -3 or m ≤ -7 :

(B) m ≤ -10 :

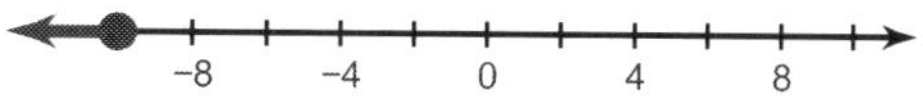

(C) m ≥ 12 or m < -12 :

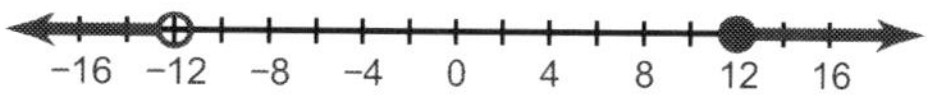

(D) m > 11 :

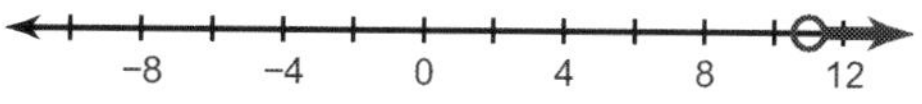

48. Which of the below represents all possible unique factors of 51

(A) 1,51

(B) 1,17,29,51

(C) 1,3,5,7

(D) 1,3,17,51

49. Find the value of x from the figure given below

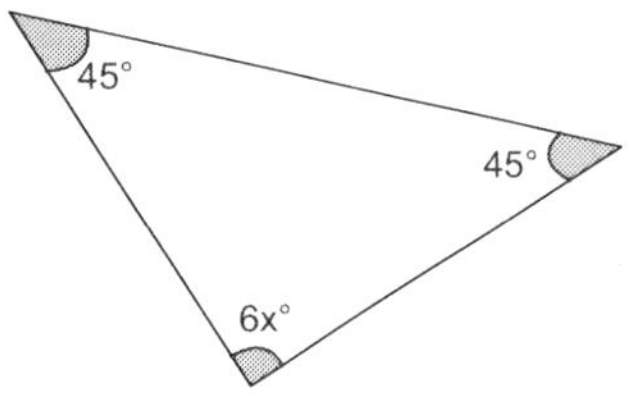

(A) 15

(B) 7

(C) 8

(D) 0

50. Find the value of x from the figure given below

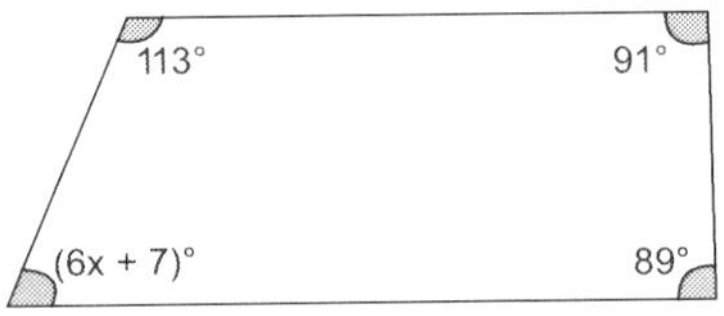

(A) 20

(B) 10

(C) 27

(D) 34

51. Find the value of x from the figure given below

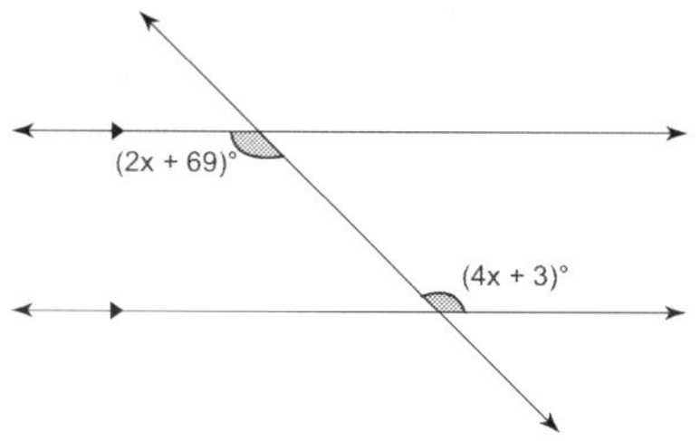

(A) 33 (B) 35

(C) 43 (D) 37

52. Solve the inequality and graph its solution $|-3x-1|+8<10$

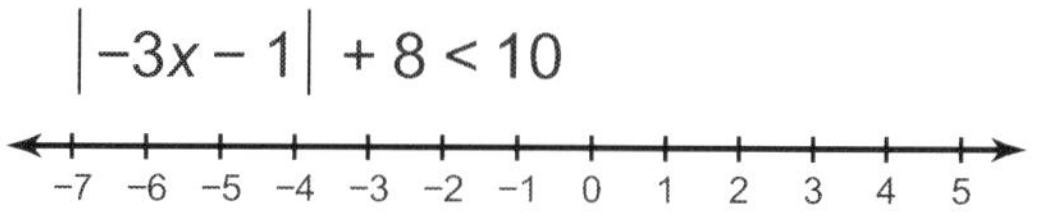

53. Find the midpoint of the line segment $\overline{AB}$ given below

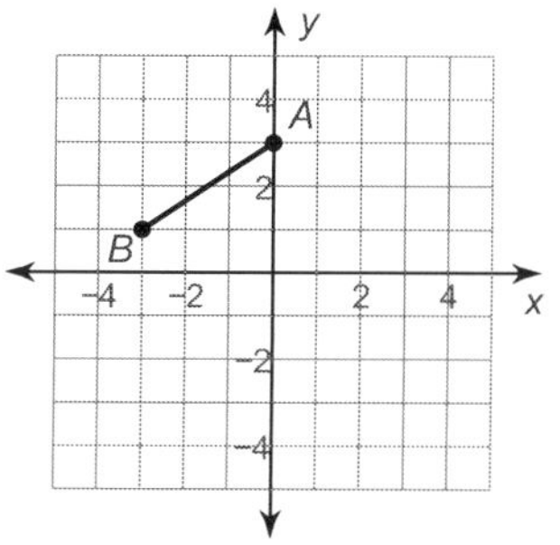

(A) (3,5) (B) (-1.5,-1)

(C) (-1,1.5) (D) (-1.5,2)

54. Stephanie reduced the size of a painting to a width of 0.9 in. What is the new height if it was originally 12 in tall and 4.5 in wide ?

(A) 60 in (B) 4.5 in

(C) 2.4 in (D) 0.2 in

55. Economists have found that the amount of corruption in a country is correlated to the productivity of that country. Productivity is measured by gross domestic product (GDP) per capita. Corruption is measured on a scale from 0 to 100 with 0 being highly corrupt and 100 being least corrupt:

Corruption Scor	GDP Per Capita ($)
19	3,580
36	8,990
48	12,200
50	13,600
61	13,000
65	23,500

This can be modeled by the equation $y = 346x - 3620$ where x is the corruption score and y is GDP per capita in dollars.

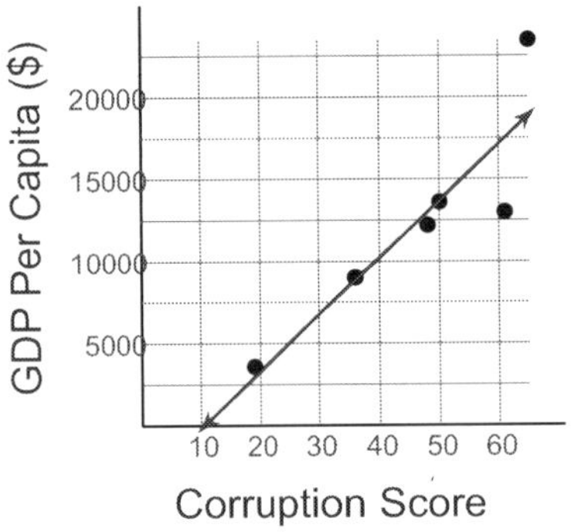

A GDP per capita of $7,800 corresponds to what corruption score, according to the model?

(A) 26

(B) 16

(C) 40

(D) 33

56. Evaluate $-\frac{8}{5} + -3\frac{1}{3} - -2$

BR High school conducted a survey on 1800 students for their favorite hobbies. The below pie chart represents percentage of student votes. Answer the questions 57 and 58.

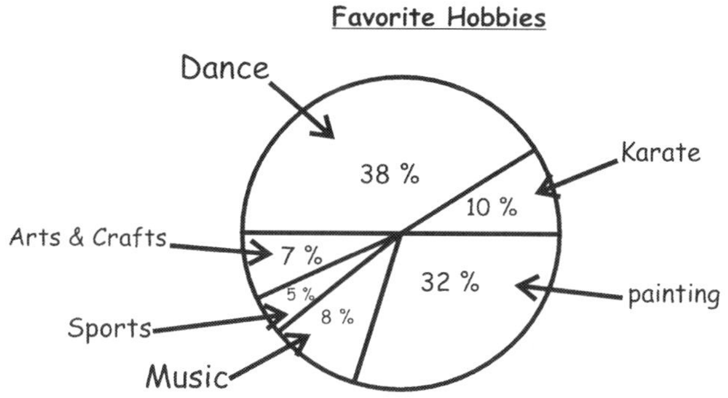

57. What percentage of students voted for music than sports ?

(A) 6% (B) 5%

(C) 4% (D) 3%

58. How many students voted for Arts & crafts and painting together ?

(A) 702 (B) 602

(C) 660 (D) 645

SHSAT Practice Test 3

1. Evaluate $|n+6| = 1$, And find the sum of the solutions

(A) 12 (B) -2

(C) -12 (D) -10

2. Find $x + y^2$ if $15x + 19y = 95$ and $2x - 19y = 228$

(A) 114 (B) 119

(C) 206 (D) 86

3. A straight line of slope $\frac{7}{9}$ is passing through the points $A(-4, -6)$ and $B(x, 1)$. Find the value of x.

(A) 6 (B) 4

(C) -4 (D) 5

4. Find the equation of the straight line perpendicular to $y = \frac{3}{2}x - 5$ and passing through the point $C(-2, 4)$

(A) $y = -\frac{5}{3}x + \frac{8}{3}$ (B) $y = -\frac{2}{3}x + \frac{8}{3}$

(C) $y = \frac{2}{3}x + \frac{8}{3}$ (D) $y = \frac{5}{3}x + \frac{8}{3}$

5. The original price of a microscope is \$624.50. If the discount offered on the microscope is 35%, what would be its selling price ?

(A) \$686.95 (B) \$218.57

(C) \$405.93 (D) \$562.05

6. Volume(V) of the rectangular box DEFG is 64 cubic ft, with length equals to 8 ft and width as 2 ft. what is the height of the rectangular box DEFG ?

(A) 1 (B) 6

(C) 3 (D) 4

7. Solve for U and T. If U + T = 29 and UT =100 where U and T are positive integers.

(A) 25 and 2 (B) 14 and 15

(C) 25 and 4 (D) 20 and 9

8. If R # P = 2R + P - 5 what is the value of $\frac{2 \# 3}{1 \# 2}$ = ?

(A) 4 (B) 2

(C) -2 (D) 1

9. Rik is running a race that is K miles long. If he has completed 17 miles distance. Which expression below shows the number of miles remaining to finish the race ?

(A) K + 17 (B) K - 17

(C) K × 17 (D) $\frac{K}{17}$

10. Dan mows the lawn in ten hours. Tom can mow the same lawn in 15 hours. Together they can complete to mow the same lawn in how many hours.

(A) 5 hours (B) 4 hours

(C) 7 hours (D) 6 hours

11. A straight line passes through the points (-2, -2) and (0, p) with slope as $\frac{9}{2}$ then p = ?

(A) 7 (B) 5

(C) -7 (D) 1

12. When 5% of tax is added to a milk can of $10.00. What is the sale price.

(A) $10.50 (B) $0.50

(C) $9.50 (D) $11.50

13. Maria had a coupon of 50% discount. She selected a dress priced at $15.96. What will be the selling price of the dress after discount.

(A) $23.92 (B) $18.34

(C) $7.90 (D) $7.98

14. Evaluate ${}_{24}C_4$ - 1

(A) 255,023 (B) 255,024

(C) 2,125 (D) 10,625

15. Evaluate $\frac{{}^{10}P_{10}}{5}$

(A) 670,580 (B) 90,720

(C) 525,430 (D) 725,760

16. The radius of the circular ground is 24m and area of inner circle 616 m^2. Find the width of circular track?

(A) 10 m (B) 12 m

(C) 15 m (D) 9 m

17. A train of length 120 meters, is moving at a speed of 30 m/s can, crosses a platform of x meters long in 20s. What is the length of the platform ?

(A) 520 m (B) 420 m

(C) 480 m (D) 410 m

18. Rita says, "If you reverse my own age, the figures represent my friend Amy's age. she is, of course, senior to me and the difference between our ages is one-eleventh of their sum." Find Rita's age?

(A) 23 years (B) 34 years

(C) 45 years (D) 30 years

19. David gets on the elevator at the 11th floor of a building and rides up at the rate of 57 floors per minute. At the same time, Albert gets on an elevator at the 51st floor of the same building and rides down at the rate of 63 floors per minute. If they continue travelling at these rates, then at which floor will their paths cross ?

(A) 19 (B) 28

(C) 30 (D) 37

20. Cathy purchased a Play Station at $219.99 and sells it at a 20% price increase.
Find the selling price of the Play Station.

(A) $175.99 (B) $252.99

(C) $230.99 (D) $263.99

21. Dora added 60 instead of 33 in her calculations. Find the percentage change in her calculations.

(A) 81.8% decrease (B) 27% increase

(C) 81.8% increase (D) 98.4% increase

22. Andy gave a personal loan to Tim an amount of $365 at 10% interest. If Tim repays it after 4 years, how much amount did Andy received ?

(A) $36.50 (B) $534.40

(C) $511.00 (D) $146.00

23. Find the missing terms in the below arithmetic sequences

..., 27, ___, ___, ___, 827, ...

(A) -171, -369, -567 (B) 227, 427, 627

(C) -173, -373, -573 (D) -172, -371, -570

24. Evaluate $(-8)^2 - ((-4) - 4)$

(A) 66 (B) 73

(C) 71 (D) 72

25. Find the missing terms in the below sequences

..., -3, ___, -48, ...

(A) 4 (B) -12

(C) 1 (D) -8

26. $$\frac{3}{r+6} = \frac{11}{r}$$

(A) -8.25 (B) 8

(C) -4.8 (D) 3.9

27. Bella mixes together 5 gallons of Brand X fruit juice and 4 gallons of apple juice. Find the percentage of fruit juice in Brand X if the mixture contains 50% fruit juice.

(A) 16% (B) 10%

(C) 6% (D) 5%

28. Working alone, John can solve a puzzle in 15 minutes. One day he worked with his friend Henry and it only took 6.67 minutes to solve the puzzle. How long does it take Henry to solve the puzzle by himself ?

(A) 9.86 minutes (B) 12.01 minutes

(C) 15.46 minutes (D) 14.92 minutes

29. Maya made a trip to her grandparents house and back. The trip there took four hours and the trip back took five hours. She averaged 24 mi/h on the return trip. Find her average speed to reach her grandparents house ?

(A) 40 mi/h (B) 45 mi/h

(C) 30 mi/h (D) 35 mi/h

30. $P(A) = \frac{9}{20}$ $P(A \text{ and } B) = \frac{27}{80}$ *then* $P(B) = ?$

(A) $\frac{1}{4}$ (B) $\frac{9}{20}$

(C) $\frac{3}{4}$ (D) $\frac{1}{5}$

31. A bag contains four red marbles, five blue marbles, and three yellow marbles. Three times, you randomly pick a marble, return it to the bag, and then mix the marbles. Find the probability of picking the first marble as red, the second marble as blue, and the third marble as yellow.

(A) $\frac{1}{4}$ (B) $\frac{2}{9}$

(C) $\frac{5}{18}$ (D) $\frac{5}{144}$

32. Rik has four white socks, six brown socks, and six black socks. Find the probability of him randomly picking two matching pair of black socks ?

(A) 0.045 (B) 0.044

(C) 0.125 (D) 0.255

33. The difference between simple interest and compound interest on \$35000 at the rate of 12 % per annum for 2 years is?

(A) \$ 782 (B) \$504

(C) \$656 (D) \$958

34. Tanya sold her book rack at a loss of 13 %. If she had sold the book rack for \$ 84 more, he would have gained 15%. Find the selling price of the book rack to get a profit of 20 %?

(A) \$360 (B) \$420

(C) \$ 480 (D) \$ 500

35 Jack reads a book of 140 pages .How many 3's are there in page numbers ?

(A) 29 (B) 31

(C) 27 (D) 34

36. A minute cell divides into 3 parts every 3 minutes. In 30 minutes how many cells are formed ?

(A) 3^9 (B) 3^{10}

(C) 3^{15} (D) 30

37. Sai runs at the rate of 5 miles per half hour. Sam runs at the rate of 1.5 times faster than his friend sai. If sams speed is Y miles per hour. What is the value of Y ?

(A) 7.5 miles per hour (B) 15 miles per hour

(C) 7.5 miles (D) 12 miles per hour

38. Maria and Kate are selling baked goods for a school fundraiser. They are selling cupcakes and sugar cookies. Mary sold 7 cupcakes and 12 sugar cookies for total \$162. Jasmine sold 10 cupcakes and 6 sugar cookies for total \$120. Find the cost a cupcake and a sugar cookie.

(A) cupcake = \$6 , sugar cookie = \$10
(B) cupcake = \$2 , sugar cookie = \$10
(C) cupcake = \$6 , sugar cookie = \$5
(D) cupcake = \$10 , sugar cookie = \$8

39. How many unique permutations of the letters in the word **DOUBLES** are possible ?

(A) 5040 (B) 5620

(C) 6020 (D) 3840

40. 100 people attended a local charity event. If each of the attendies shake hands with everyone else how many hand shakes are done ?

(A) 990 (B) 4920

(C) 3650 (D) 4950

41. Amy gave a personal loan to Sara an amount of $1,700 at 4% interest , compounded annually. If Sara repays it after 3 years, how much amount will Amy receive ?

(A) $1,904.00 (B) $1,912.27

(C) $1,919.26 (D) $1,905.06

42. There are 15 animals in the barn. Some are chickens and some are sheep. There are 50 legs in all. How many of each animal are there ? How many chickens are there in the barn ?

(A) 5 chickens (B) 12 chickens

(C) 13 chickens (D) 11 chickens

43. Two machines X and Y takes 2 inputs and produce one output which is greatest of all. Outputs of machines X and Y become inputs to machine Z and the greatest of all becomes the output. Based on the given information and the below figure what is the final output produced ?

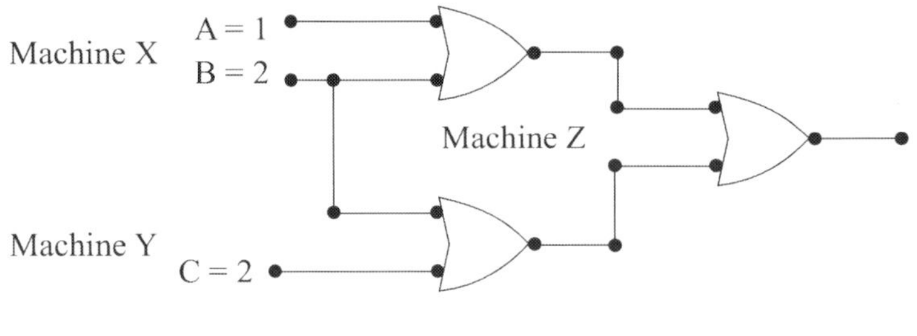

(A) 1 (B) 4

(C) 3 (D) 2

44. Simplify the expression 9(x + 2) - 7(3 + 2x)

(A) -10 (B) -7x + 6

(C) -5x + 6 (D) -5x - 3

45. Simplify the expression -18x - 16(x - 15)

(A) -331 + 340x (B) -46x + 240

(C) -314 + 340x (D) -34x + 240

46. Simplify the expression and plot the graph $\frac{11 + x}{18} < 1$

(A) x < 7 :

(B) x < 5 :

(C) x< -4 :

(D) x > 7 :

47. Simplify the inequality and graph the solution

-6r + 7 > 61 and 8 -3r ≤ 38

(A) r > -6 :
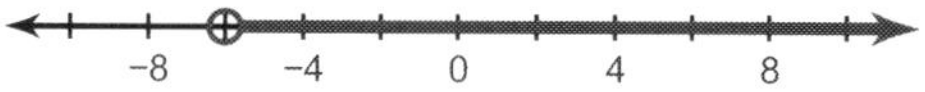

(B) -6 < r ≤ 5 :
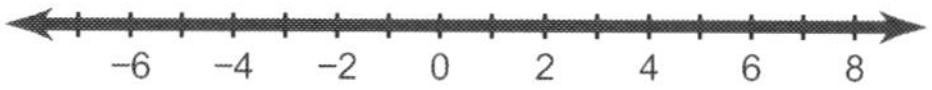

(C) r ≥ -1 :
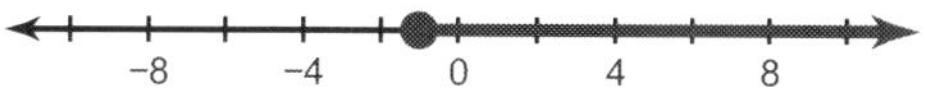

(D) -10 ≤ r < -9 :
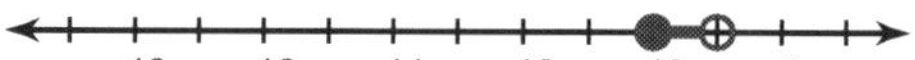

48. Which of the below represents all possible unique factors of 62

(A) 1,2,5,62 (B) 1,2,31,62

(C) 1,62 (D) 1,31,62

49. Find the value of x from the figure given below

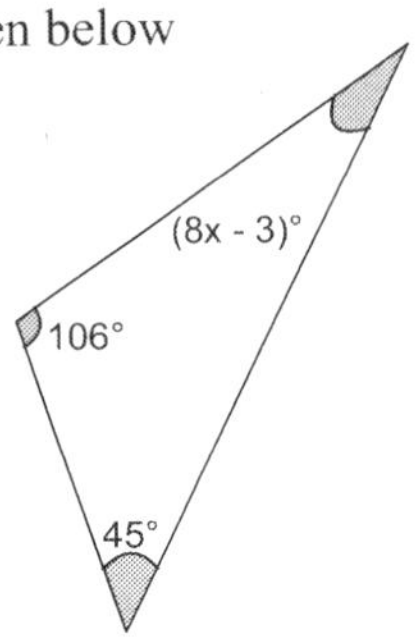

(A) 4 (B) 3

(C) 1 (D) 5

50. Find the value of x from the figure given below

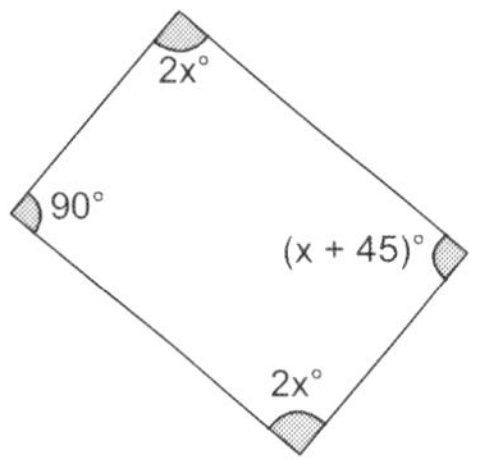

(A) 36 (B) 45

(C) 29 (D) 22

51. Find the value of x from the figure given below

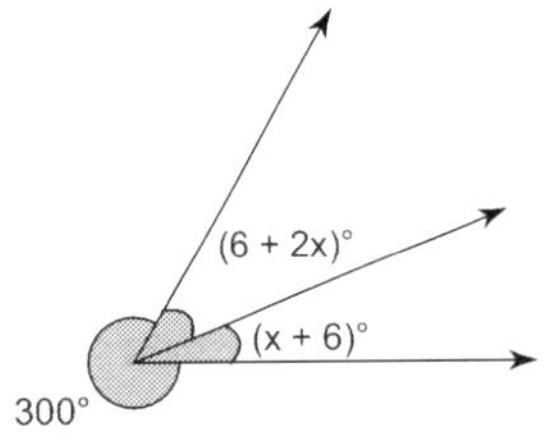

(A) 17 (B) 35

(C) 41 (D) 16

52. Solve the inequality and graph its solution

$$6|3+8p| \leq 30$$

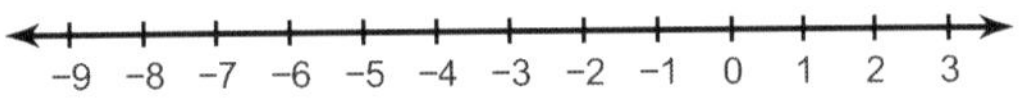

53. Find the midpoint of the line segment $\overline{MN}$ given below

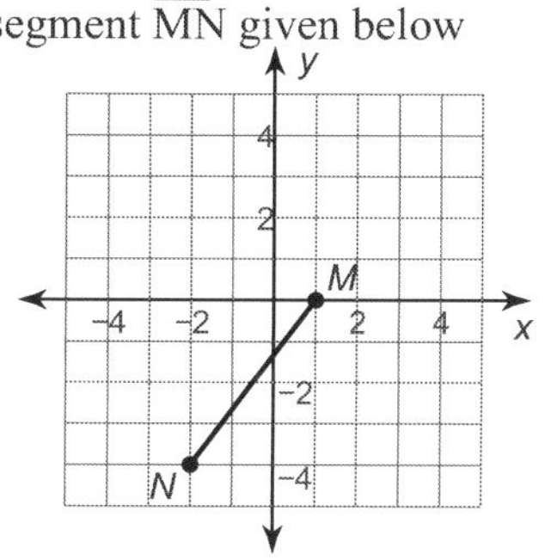

(A) (-0.5,2) (B) (-3,-0.5)

(C) (4,4) (D) (-1.5,-2)

54. Heather reduced the size of a rectangle to a width of 1 in. What is the new height if it was originally 3 in wide and 7.5 in tall?

(A) 2 in (B) 2.5 in

(C) 2.2 in (D) 0.3 in

55. Krystal collects coins. Over a three-year period she collected 1,000 nickels. After organizing them by year, she found that the number of nickels from a given year was related to the number minted that year:

Minted (100 mill.)	In Sample
3.5	3
9.6	18
10.7	25
13.3	21
15.4	35
16.6	33

Krystal discovered that this can be modeled by the equation $y = 2.33x - 4.29$ where x is the number of nickle minted in a particular year in hundreds of millions and y is the number of nickels from that year in her sample.

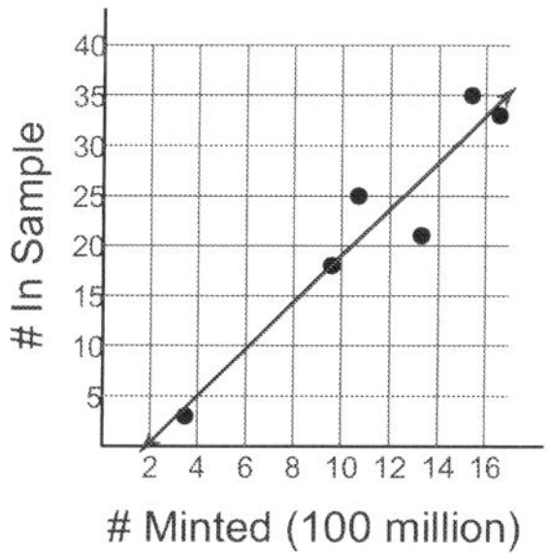

A sample of 1,000 nickels contains 10 from a particular year. According to the model, how many would you expect to have been minted that year? Round your answer to the nearest tenth of a hundred million.

(A) 4.6 hundred million nickels (B) 6.8 hundred million nickels

(C) 6.1 hundred million nickels (D) 8 hundred million nickels

56. Evaluate $\frac{8}{7} + -\frac{1}{2} - 2\frac{1}{3}$

BR High school conducted a survey on 1800 students for their favorite hobbies. The below pie chart represents percentage of student votes. Answer the questions 57 and 58.

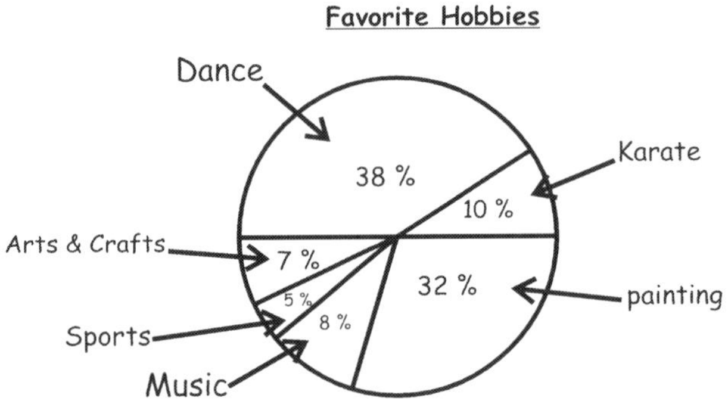

57. By what percent is the number of students who voted for Arts & crafts and dancing together are more than those who voted for painting and music together ?

(A) 14.5% (B) 5%

(C) 9% (D) 8%

58. Find the ratio between the number of students who voted for dancing and karate than who voted for painting ?

(A) 2:3 (B) 1:3

(C) 3:1 (D) 3:2

SHSAT Practice Test 4

1. Evaluate $y(y + y + z + 3)$; where $y = 7$, and $z = -8$

(A) 55 (B) 63

(C) 70 (D) 69

2. Evaluate $|-6 + 7m| = 50$, and find the sum of the solutions

(A) $\frac{1}{2}$ (B) 8

(C) $\frac{11}{3}$ (D) $\frac{12}{17}$

3. If $36 \cdot 6^{-3r-1} = 6^{-2r}$, find the value of $r^2 + 5r$?

(A) -4 (B) 2

(C) 6 (D) 5

4. Find $(x + y)^2$ if $4x + 17y = 119$ and $26x + 17y = -255$

(A) 18 (B) 36

(C) 34 (D) 289

5. A straight line of slope $-\frac{1}{5}$ is passing through the points $A(-8, 9)$ and $B(7, y)$. Find the value of y.

(A) 9 (B) 8

(C) -2 (D) 6

6. Find the equation of the straight line perpendicular to $y = -\frac{2}{7}x + 4$ and passing through the point D(3, 4)

(A) $y = \frac{7}{2}x - \frac{13}{2}$ (B) $y = \frac{13}{2}x + \frac{7}{2}$

(C) $y = x + \frac{7}{2}$ (D) $y = -\frac{13}{2}x + \frac{7}{2}$

7. Brand X of a vacuum cleaner is sold at a price of \$109.95. A store is offering a 40% discount today. Find the selling price of the vacuum cleaner.

(A) \$131.94 (B) \$93.46

(C) \$65.97 (D) \$43.98

8. Find the volume of a square prism measuring 7 ft along each edge of the base and 8 ft tall. Answer to the nearest tenth

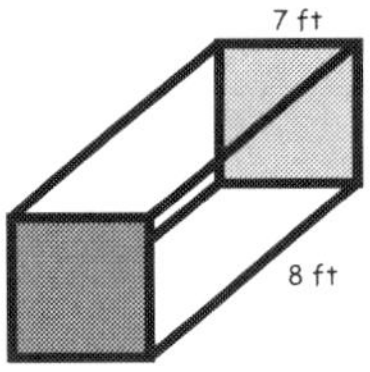

(A) 392 ft^3 (B) 593.4 ft^3

(C) 345 ft^3 (D) 690 ft^3

9. Which of the below options form a right angle triangle ?

(A) 21 , 28 , 34 (B) 14 , 48 , 49

(C) 39 , 80 , 87 (D) 12 , 35 , 37

10. Which of the below options describes the transformation of the below figure XYZ ?

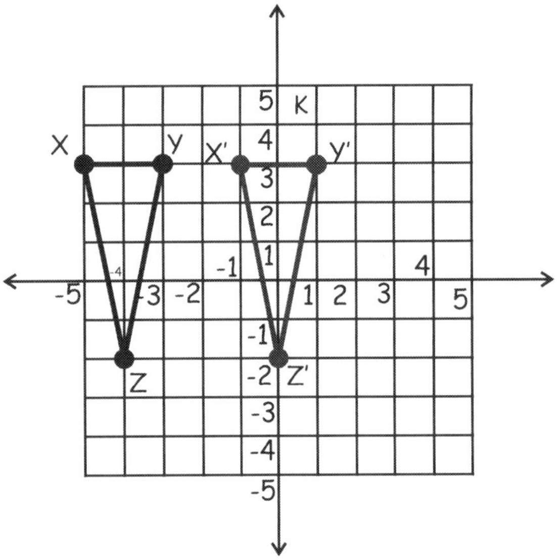

(A) translation of 6 units to right and 2 units up

(B) translation of 2 units right and 1 unit up

(C) rotation across x = -2

(D) translation of 4 units right

11. Find the volume of a cone with a diameter of 16 cm and a height of 20 cm.
Answer to the nearest tenth.

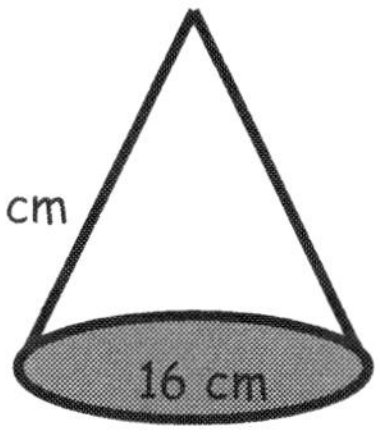

(A) 1340.4 cm^3 (B) 1139.4 cm^3

(C) 649.5 cm^3 (D) 569.7 cm^3

12. Monopoly game costs \$25.00. The store is offering a discount of 30% on it. What is the selling price of the game ?

(A) \$26.25 (B) \$7.50

(C) \$32.50 (D) \$17.50

13. Evaluate $\frac{{}^{11}P_6}{9} = ?$

(A) 30,640 (B) 32,924

(C) 36,960 (D) 33,241

14. Working alone , Jack pick forty bushels of apples in 11 hours. Julia can pick the same amount in 9 hours.
Find how long it would take them if they worked together.

(A) 4.45 hours (B) 5.9 hours

(C) 3.89 hours (D) 4.95 hours

15. Evaluate $\frac{{}^{13}C_7}{11} = ?$

(A) 128 (B) 844

(C) 240 (D) 156

16. A monkey climbs 30 feet at the beginning of each hour and rests for a while and slips back 20 feet. He starts back again at the beginning of the next hour. If he starts climbing at 8.00 a.m., at what time will he reach 120 feet from the ground?

(A) 4 p.m. (B) 5 p.m.

(C) 6 p.m. (D) 3 a.m

17. Diameter of a circle is 3/5th of sum of length and breadth of rectangle. The perimeter of rectangles is 60 m and length of rectangle is 1.5 times of breath of rectangle. Find the area of circle?

(A) 306.5 m^2 (B) 216.5 m^2

(C) 254.5 m^2 (D) 297.5 m^2

18. Find the number of unique permutations of the word **EQUINOX** can be formed ?

(A) 7,540 (B) 4,280

(C) 6,120 (D) 5,040

19. Kim purchased a Monopoly game for $19.95 and sold at 90% price hike. What is the selling price of the monopoly game ?

(A) $17.95 (B) $37.91

(C) $37.80 (D) $1.99

20. The original price of a car is $37,000.00 and the tax need to be paid for the car is 6%, what is the selling price of the car ?

(A) $2,220.00 (B) $39,220.00

(C) $34,780.00 (D) $42,550.00

21. Ella added 66 in her calculations instead of 81.7. Find the percentage change in her calculations.

(A) 23.8% increase (B) 19.2% decrease

(C) 15.7% increase (D) 80.8% decrease

22. Lola gave a personal loan to Rosy an amount of $1,330 at 6% interest. If Rosy repays it after 8 years, how much amount did Lola received ?

(A) $2,119.82 (B) $1,968.40

(C) $1,409.80 (D) $638.40

23. Find the missing terms in the below arithmetic sequences

..., 29, ___, ___, 23, ...

(A) 25, 23
(B) 21, 18
(C) 27, 25
(D) 24, 21

24. Find the missing terms in the below sequences

..., -35, ___, ___, ___, ___, ___, -65, ...

(A) -30, -27, -24, -21, -18
(B) -32, -29, -26, -23, -20
(C) -40, -45, -50, -55, -60
(D) -30, -25, -20, -15, -10

25. Evaluate $-\dfrac{5 \times 2 \times 2}{4}$

(A) -13
(B) -3
(C) -5
(D) -1

26. Solve the proportion $\dfrac{13}{5} = \dfrac{a+1}{a}$

(A) 0.63
(B) -7.56
(C) -13
(D) 7

27. Liam mixes 8 L of Brand A fruit juice and 4 L of Brand B fruit juice which contains 38% orange juice. Find the percent of orange juice in Brand A if the mixture contained 36% orange juice.

(A) 55%
(B) 35%
(C) 19%
(D) 10%

28. Working alone, Ella can build a bird cage in 15 hours. One day her friend Rocky helped her and it only took 8.57 hours. How long does it take Rocky to build it alone ?

(A) 19.99 hours
(B) 15.23 hours
(C) 14.93 hours
(D) 16.34 hours

29. If $P(B) = \frac{2}{5}$ $P(A \text{ and } B) = \frac{3}{25}$ *then* $P(A) = ?$

(A) $\frac{7}{10}$

(B) $\frac{77}{400}$

(C) $\frac{7}{20}$

(D) $\frac{3}{10}$

30. There are six nickels and six dimes in your pocket. Three times, you randomly pick a coin out of your pocket, return it to your pocket, and then mix-up the change in your pocket. All three times, the coin is a nickel.

(A) $\frac{3}{11}$

(B) $\frac{24}{91}$

(C) $\frac{1}{8}$

(D) $\frac{36}{1331}$

31. Marked Price of a furniture set gives 30% of profit in its cost price. if store gives 10%, 15% successive discount on the furniture set, and it is sold for \$1989. Find the selling price of the furniture set if he sells at 115% of original cost price?

(A) \$ 1750

(B) \$ 2250

(C) \$ 2000

(D) \$ 2300

32. Two machines X and Y takes 2 inputs and produce one output which is greatest of all. The outputs of machines X and Y become inputs to machine Z and get added to produce the output. Based on the given information and the below figure find the missing input value B ?

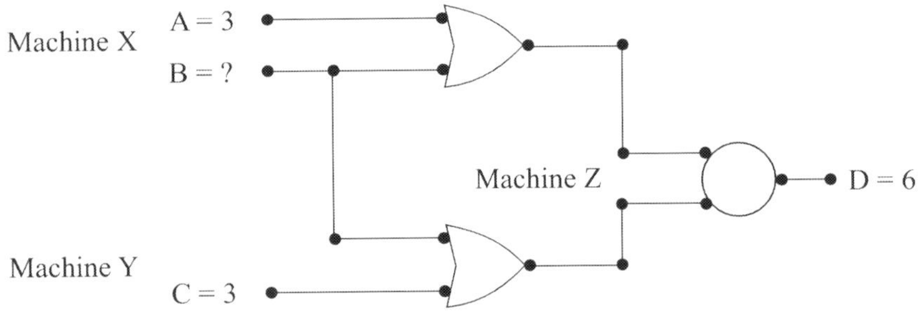

(A) 6

(B) 3

(C) 2

(D) Cannot be determined

33. A box contains 11 Skittles , 4 Sneakers , 10 Twigs , 12 Kit-Kats , 13 M & M's. Harper picks an item from the box , what is the probability that it would be a M & M or a Kit-Kat ?

(A) 25 %

(B) 50 %

(C) 75 %

(D) 55 %

34. Tom completes his walking on circular ground X by 36 minutes and Mary completes her walking on another circular ground Y by 90 minutes. The total length of Ground X is 3/5 times of the total length of Ground Y. Find the speed ratio of Mary to Tom ?

(A) 2:3 (B) 3:2

(C) 4:1 (D) 1:4

35. Find the missing letter from the below series

A , H , F , M , K , R , P , ?

(A) D (B) W

(C) L (D) X

36. The base of a triangular sign board is 23 feet with a height of 17 feet. Find the cost of painting it at the rate of \$2 per square feet.

(A) \$391 (B) \$195.5

(C) \$198 (D) \$401

37. From a square sheet of side 14 inches a circular sheet with maximum dimensions is taken away. Find the area of the remaining sheet.

(Note : $\prod = \frac{22}{7}$)

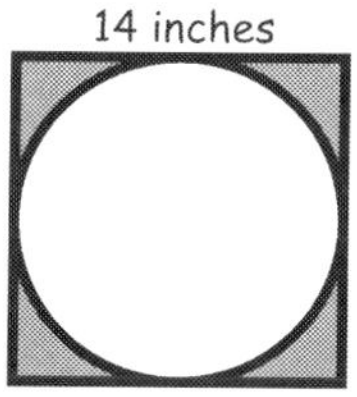

(A) 154 in^2 (B) 196 in^2

(C) 64 in^2 (D) 42 in^2

38. How many unique permutations of the letters in the word **SPECTRUM** can be formed ?

(A) 40,220 (B) 40,320

(C) 38,450 (D) 42,580

39. When the integer n is divided by 17, the quotient is x and the reminder is 5. When n is divided by 23, the quotient is x and the reminder is 14. Which of the following is true ?

(A) $23x + 17y = 19$ (B) $17x - 23y = 9$

(C) $17x + 23y = 19$ (D) $14x + 5y = 6$

40. Ken gave a personal loan to Ben an amount of \$5,800 at 8% interest. compounded annually. If Ben repays it after 6 years, how much amount will Ken receive ?

(A) \$8,584.00 (B) \$9,203.87

(C) \$9,214.91 (D) \$9,189.22

41. There are 15 animals in the barn. Some are chickens and some are sheep. There are 50 legs in all. How many of each animal are there ? How many sheeps are there in the barn ?

(A) 10 sheeps (B) 3 sheeps

(C) 2 sheeps (D) 4 sheeps

42. Ship X left Port A and traveled to reach Port B. Ship Y left eight hours later traveling 12 mph faster in an effort to catch up to it. After six hours Ship Y speed is caught up to Ship X. What was the average speed of Ship X ?

(A) 13 mph (B) 11 mph

(C) 8 mph (D) 9 mph

43. There are six boys and four girls in a class. The teacher randomly selects three different students to answer questions. Find the probability of teacher choosing the first student a boy, the second and third students as girls.

(A) 0.1 (B) 0.074

(C) 0.25 (D) 0.048

44. Simplify the expression 4(n -9) - (n + 3)

(A) 16n - 39 (B) 10n - 39

(C) 11n - 39 (D) 3n - 39

45. Simplify the expression 18(1 + 3p) + 7

(A) 71 + 63p (B) 25 + 54p

(C) 50 + 63p (D) 53 + 63p

46. Simplify the expression and plot the graph $1 > \frac{m}{8} - 1$

(A) m<16 :
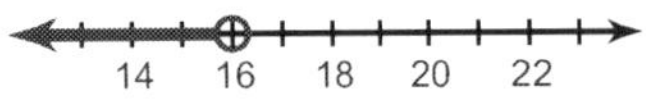

(B) m<-42 :
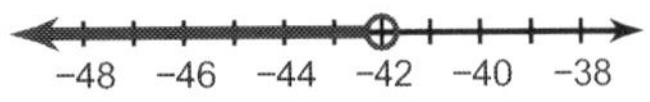

(C) m<-34 :

(D) m<-21 :
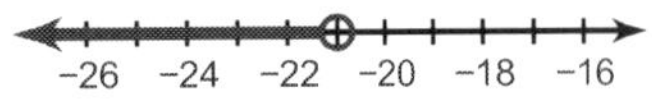

47. Simplify the inequality and graph the solution

$$6x - 11 \geq 43 \text{ or } 6x - 4 < 14$$

(A) $x > 6$ or $x < 1$: 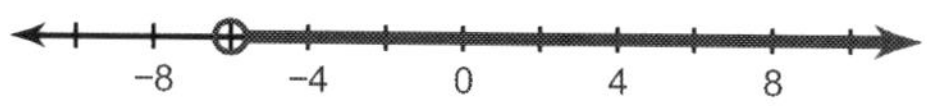

(B) $x \geq 11$: 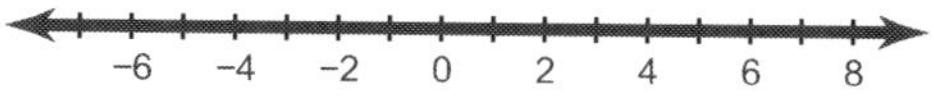

(C) $x \geq 9$ or $x < 3$: 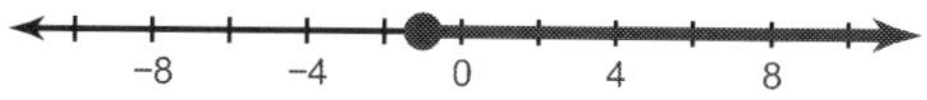

(D) $x \geq 9$: 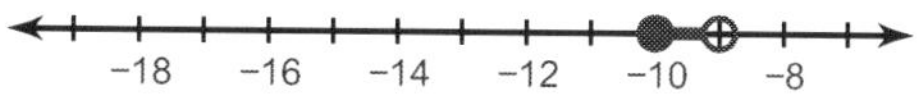

48. Which of the below represents all possible unique factors of 45

(A) 1,3,5,9,15,45

(B) 1,3,5,9,15

(C) 1,3,45

(D) 1,3,15,45

49. Find the value of x from the figure given below

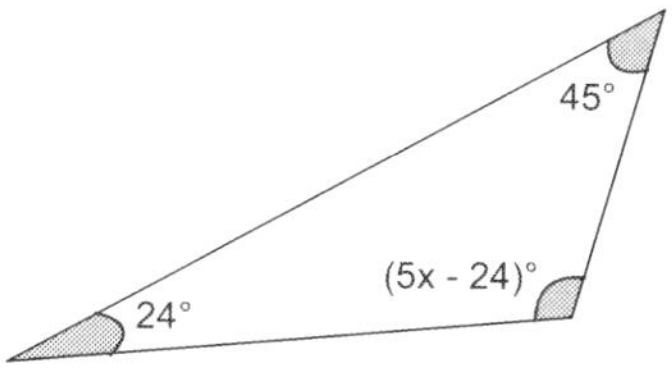

(A) 22

(B) 18

(C) 13

(D) 27

50. Find the value of x from the figure given below

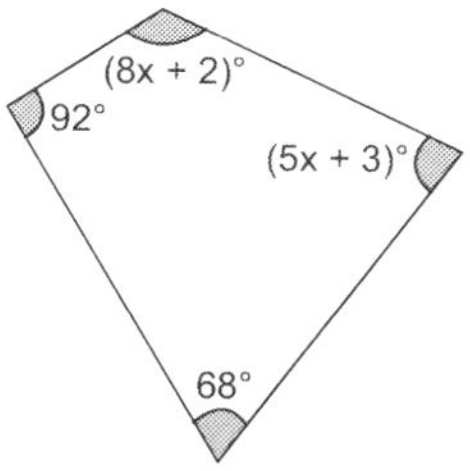

(A) 24

(B) 41

(C) 32

(D) 15

51. Find the value of x from the figure given below

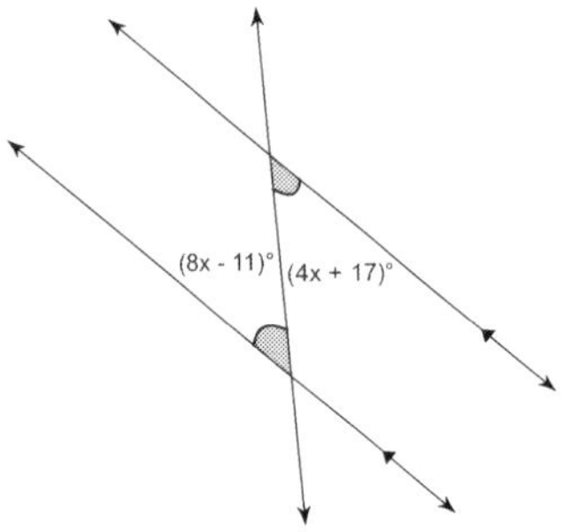

(A) 4 (B) 7

(C) 2 (D) 9

52. Solve the inequality and graph its solution

$$|-7b - 9| + 7 > 72$$

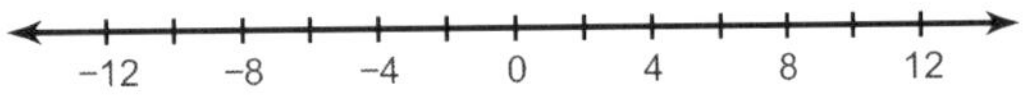

53. Find the midpoint of the line segment $\overline{PQ}$ given below

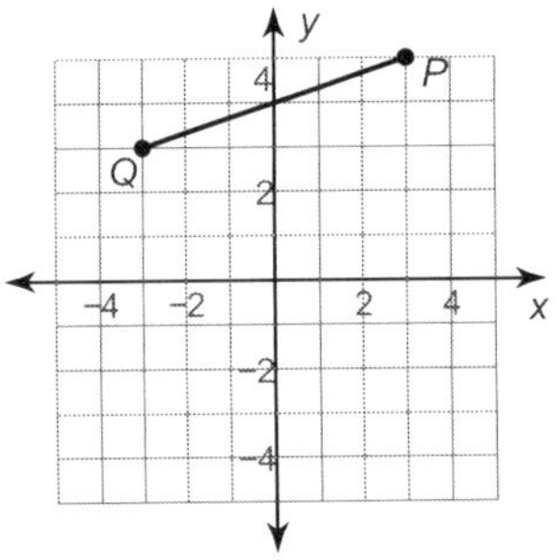

(A) (-9,1) (B) (4,0)

(C) (3,1) (D) (0,4)

54. Lisa reduced the size of a rectangle to a height of 0.8 in. What is the new width if it was originally 1.6 in tall and 3.2 in wide?

(A) 0.4 in (B) 1.6 in

(C) 1.4 in (D) 1.2 in

55. The height and weight of several adults were recorded:

Height (ft)	Weight (lbs)
4.55	95.5
5.35	129
5.65	145
5.95	165
6.2	179
6.35	184

It was discovered that this can be modeled by the equation $y = 50.7x - 138$ where x is height in feet and y is weight in pounds

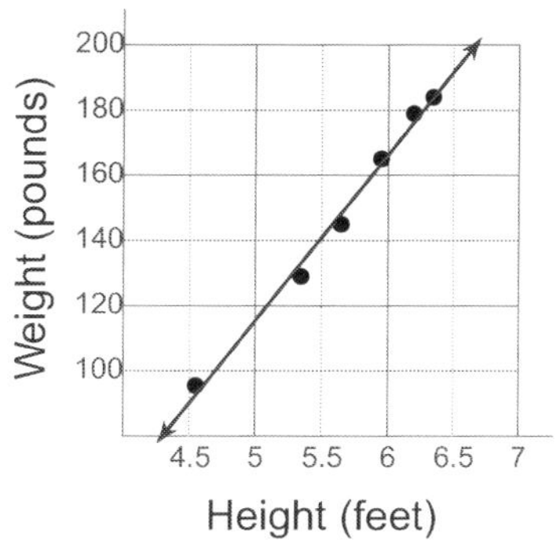

What height corresponds to a weight of 110 pounds? Round your answer to the nearest hundredth

(A) 4.14 ft

(B) 5.69 ft

(C) 4.89 ft

(D) 4.02 ft

56. Evaluate $-\frac{4}{3} - 0 + \frac{4}{3}$

SS High school conducted a survey on 6800 students for their favorite Soft Drinks. The below pie chart represents percentage of student votes. Answer the questions 7 - 12.

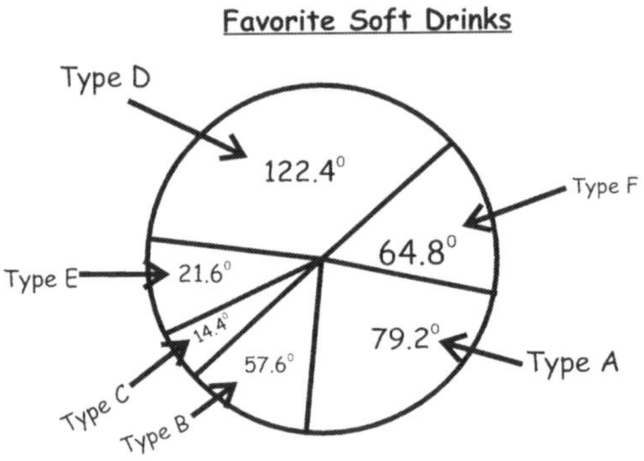

57. What is the difference between the total number of votes for types A and C together and who preferred types D and F together ?

(A) 959 (B) 960

(C) 952 (D) 953

58. Find the ratio of the number of student votes for type F and type A soft drinks

(A) 9:11 (B) 3:13

(C) 6:11 (D) 5:11

SHSAT Practice Test 5

1. Evaluate $|6m - 7| = 43$, and find the sum of the solutions.

(A) $2\frac{1}{3}$ (B) $\frac{2}{5}$

(C) $\frac{1}{2}$ (D) $\frac{25}{3}$

2. If $3^k \cdot 3^{-35} = 3^{2k}$ then $2k + 100 = ?$

(A) 70 (B) 30

(C) -35 (D) −70

3. Find $(x - y)^2$ if $7x + 19y = -285$ and $15x - 19y = -133$

(A) -1 (B) -15

(C) 11 (D) 121

4. A straight line of slope $\frac{7}{3}$ is passing through the points $A(-5, 4)$ and $B(x, -3)$. Find the value of x.

(A) 2 (B) 5

(C) -9 (D) -8

5. Find the equation of the straight line perpendicular to $y = -4$ and passing through the point $E(-5, -5)$

(A) $y = 5$ (B) $x = -5$

(C) $x = 5$ (D) $y = \frac{5}{3}$

6. Maria purchased a dining table set for \$64.95. Find out the selling price if 2% tax is added to the purchase of the dining table ?

(A) \$61.70 (B) \$63.65

(C) \$1.30 (D) \$66.25

7. A cell phone cover has two flaps side. One flap is of the length 8.92×10^{-6} and the other is 1.60×10^{-6} What is the difference in the lengths of flaps ?

(A) 7.32×10^{-6} (B) 0.732×10^{-6}

(C) 0.0732×10^{-6} (D) 73.2×10^{-6}

8. The function f(y) = 35 + 15 h represents the amount of money, in dollars. where h is the no of hours Shan worked. He works 10 hours this week. How much money did he earn this week ?

(A) 165 (B) 225

(C) 185 (D) 180

9. Which of the below graph best represents the function

$$f(x) = x^2 + 5x + 4$$

(A)

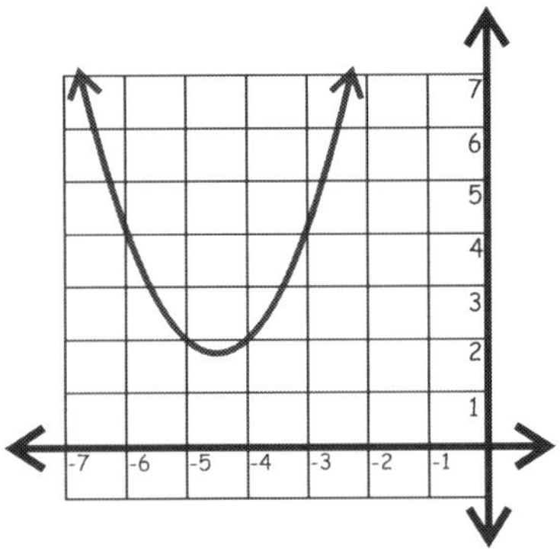

(B)

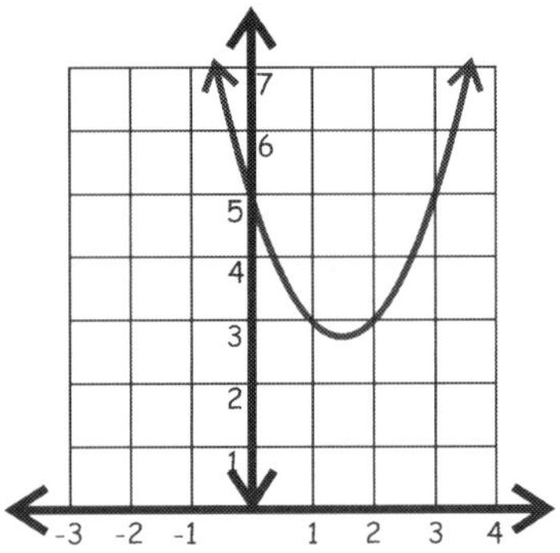

(C)

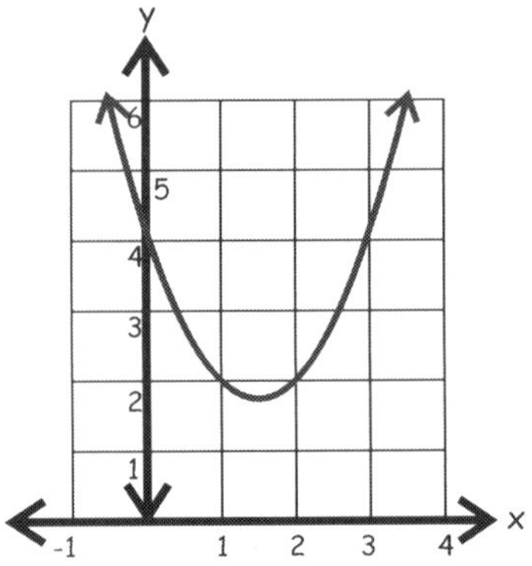

(D)

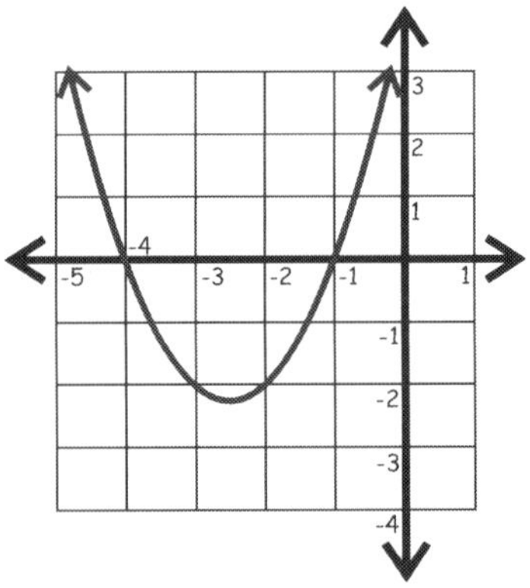

10. Dora solved puzzles each day over spring break. The number of puzzles she solved in seven days is shown.

13 , 11 , 13 , 15 , 10 , 13 , 9

She then solved 15 more puzzles on last day. How are the mean and median affected ?

(A) The mean increased and the median remained the same.

(B) The median increased and the mean remained the same.

(C) The median and the mean both remained the same.

(D) The mean and the median both decreased.

11. Find the sum of the below matrices ?

$$\begin{vmatrix} 5 & 7 \\ -2 & -4 \\ 1 & 6 \end{vmatrix} + \begin{vmatrix} 11 & 9 \\ 15 & 8 \\ 3 & -1 \end{vmatrix}$$

(A) $\begin{vmatrix} 15 & -5 \\ 10 & 20 \\ -3 & -1 \end{vmatrix}$

(B) $\begin{vmatrix} 16 & 16 \\ 13 & 4 \\ 4 & 5 \end{vmatrix}$

(C) $\begin{vmatrix} 7 & 17 \\ -8 & 9 \\ 5 & -4 \end{vmatrix}$

(D) $\begin{vmatrix} 10 & -12 \\ 16 & -6 \\ 2 & -11 \end{vmatrix}$

12. Evaluate the below matrices

$$\begin{vmatrix} -6 & 1 \\ 4 & 3 \\ 0 & 5 \end{vmatrix} - \begin{vmatrix} 9 & 3 \\ -2 & 0 \\ 3 & 1 \end{vmatrix} = ?$$

(A) $\begin{vmatrix} 15 & 8 \\ 5 & 4 \\ 9 & 7 \end{vmatrix}$

(B) $\begin{vmatrix} -15 & -2 \\ 6 & 3 \\ -3 & 4 \end{vmatrix}$

(C) $\begin{vmatrix} 14 & 7 \\ 2 & -13 \\ 15 & -21 \end{vmatrix}$

(D) $\begin{vmatrix} -9 & 15 \\ 7 & 17 \\ 15 & -1 \end{vmatrix}$

13. The function f(y) = (5/3) m, represents the amount of muffins baked by Beth. Where m is the number of muffins she gave to her friends. If she shares 24 muffins with her friends, How many muffins did she make originally ?

(A) 44 (B) 38

(C) 25 (D) 40

14. If the thickness of a pile of 80 card boards is 117.20 mm (milli meters) , how many card boards will there be in a pile which is 703.20 mm thick ?

(A) 488 (B) 480

(C) 670 (D) 467

15. Find the average of all the prime numbers between 30 and 50 ?

(A) 39.8 (B) 42

(C) 35.5 (D) 29.4

16. An Intercity bus charges are comprised of a fixed charge, together with the charge of the distance covered. For a journey of 16 miles, the charges paid are \$15.6 and for a journey of 24 miles, the charges paid are \$20.4. What will a person have to pay for traveling a distance of 30 miles ?

(A) \$ 20 (B) \$ 24

(C) \$ 48 (D) \$ 12

17. Mary makes a profit of \$1 on the first day at her new stores. On every subsequent day, she earns an income which is just double of the previous day. One the 10th day of business, her income is

(A) \$ 2^9 (B) \$ 2^{10}

(C) \$18 (D) \$ 102

18. Mia wants to go to concert and the tickets are priced at \$184.50. Her friend gave her a coupon of 35% on the concert tickets. How much does the ticket costs after the discount ?

(A) \$119.93 (B) \$166.05

(C) \$118.92 (D) \$64.58

19. Bryan purchased an Apartment at \$54,000.00 and sells it at a 40% price hike. Find the selling price of the Apartment.

(A) \$21,600.00 (B) \$62,100.00

(C) \$75,600.00 (D) \$27,000.00

20. Stella added 63 in her accounts instead of 82. What is the percentage change ?

(A) 19% decrease
(B) 30.2% increase
(C) 19% increase
(D) 23.2% decrease

21. Henry took a loan of $1,320 at 12% interest for 2 years. Find the amount he needs to pay at the end of the loan term.

(A) $1,478.40
(B) $1,636.80
(C) $316.80
(D) $1,655.81

22. A farmhouse shelters 23 animals. Some are pigs and some are chickens. Altogether there are 84 legs. How many of each animal are there ? How many chickens are there in farmhouse ?

(A) 4 chickens
(B) 20 chickens
(C) 21 chickens
(D) 8 chickens

23. Find the missing terms in the below arithmetic sequences

..., -31, ___, ___, -58, ...

(A) -22, -13
(B) -24, -15
(C) -23, -13
(D) -40, -49

24. Evaluate $\left(-\frac{8}{4}\right)-((-3)+8)$

(A) 2
(B) -8
(C) -12
(D) -7

25. Find the missing terms in the below sequences

..., 1, ___, 36, ...

(A) $-\frac{7}{3}$
(B) 6
(C) 1
(D) $-\frac{1}{3}$

26. Solve the proportion $\frac{n+6}{n} = \frac{5}{12}$

(A) 13
(B) 10
(C) -10.29
(D) -3

27. Ben wants to make 14 oz. of alloy containing 83% copper. He is planning on melting and combine metal X containing 89% copper with metal Y containing 75% copper. How much quantity of metal X and metal Y are needed to create alloy ?

(A) 8 oz. of 89% copper, 6 oz. of 75% copper
(B) 4 oz. of 89% copper, 10 oz. of 75% copper
(C) 5 oz. of 89% copper, 9 oz. of 75% copper
(D) 10 oz. of 89% copper, 4 oz. of 75% copper

28. Cathy can water the plants in 15 minutes. Macy can do the same in 11 minutes How long will it take them if they worked together ?

(A) 6.32 minutes
(B) 6.79 minutes
(C) 6.35 minutes
(D) 5 minutes

29. A train X left Virginia and traveled to Newyork . Another train Y left one hour later traveling 7 km/h faster in an effort to catch up to train X. After ten hours train Y is finally caught up. Find the average speed of train X ?

(A) 70 km/h
(B) 40 km/h
(C) 90 km/h
(D) 120 km/h

30. A vending machine contains thirteen bottles of sports drinks : four lemon-lime flavored, four orange flavored, and five fruit-punch flavored. Three times, Andy randomly grabs a bottle, return the bottle to the cooler, and then mix up the bottles. Find the probability of him picking up the first time, a lemon-lime drink. The second and third times, a fruit-punch.

(A) $\frac{1}{20}$
(B) $\frac{1}{4}$
(C) $\frac{1}{12}$
(D) $\frac{100}{2197}$

31. If $P(A) = \frac{3}{5}$ $P(B) = \frac{9}{20}$ then P(A and B) = ?

(A) $\frac{27}{80}$
(B) $\frac{27}{100}$
(C) $\frac{27}{200}$
(D) $\frac{7}{80}$

32. There are seven nickels and six dimes in Zara's bag. She randomly picks a coin from her bag and gives it to her sister. She randomly picks another coin and give it off. Find the probability that the first coin is a nickle and second is a dime ?

(A) $\frac{1}{4}$ (B) $\frac{1}{192}$

(C) $\frac{1}{128}$ (D) $\frac{7}{26}$

33. A printer numbers the pages of a story book starts with 1 and uses 3189 digits in all. How many pages does the book have?

(A) 1000 (B) 1074

(C) 1075 (D) 1080

34. Dan and John's salaries are in the ratio 8: 9 and their expenditures are in the ratio of 33: 37. If Dan saves \$ 14000 and John saves \$ 16000, then find the income of Dan?

(A) \$ 35000 (B) \$ 50000

(C) \$ 80000 (D) \$ 60000

35. A handbag was sold at a discount of 25 % and there was a profit of 20 %. If the profit earned was \$ 100 less than the discount offered, then find the selling price of the handbag ?

(A) \$ 650 (B) \$ 700

(C) \$ 750 (D) \$ 600

36. The table below shows math test scores of grade 7 students

75 , 62 , 70 , 60 , 62

Miss Rosy, the math teacher missed to add students score. If the median and mode of the data is same. what is the missing score ?

(A) 62 (B) 55

(C) 75 (D) 60

37. Find the number of terms in the arithmetic series given below

$$a_1 = -5 , a_n = -75 , s_n = -600$$

(A) 19 (B) 16

(C) 13 (D) 15

38. Find the number of terms in the series given below

$$\sum_{i=1}^{n} (6i - 13) = 525$$

(A) 16 (B) 15

(C) 18 (D) 11

39. Find the number of terms in the series given below

$$\sum_{i=1}^{7} \left(\frac{-1}{3}\right)^{i-1}$$

(A) $\frac{547}{729}$ (B) $\frac{547}{726}$

(C) $\frac{547}{728}$ (D) $\frac{3}{4}$

40. Find the number of terms in the below series

$$a_1 = 2 , r = -5 , s_n = 42$$

(A) 7 (B) 5

(C) 6 (D) 3

41. In how many unique ways the letters of the word **BRIGHTLY** be arranged ?

(A) 40,320 (B) 38,470

(C) 28,320 (D) 40,240

42. The figure below shows the dimensions of isosceles triangle in terms of x. Find the area of the triangle ?

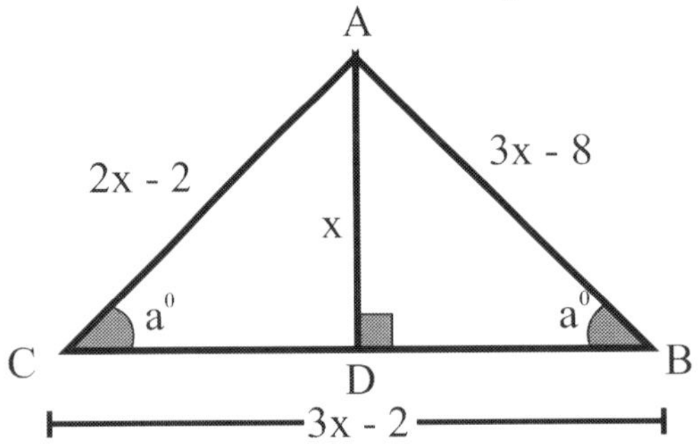

(A) 24 (B) 30

(C) 48 (D) 60

43. Two machines X and Y takes 2 inputs and produce one output which is product of all. Based on the given information and the below figure find the missing value ?

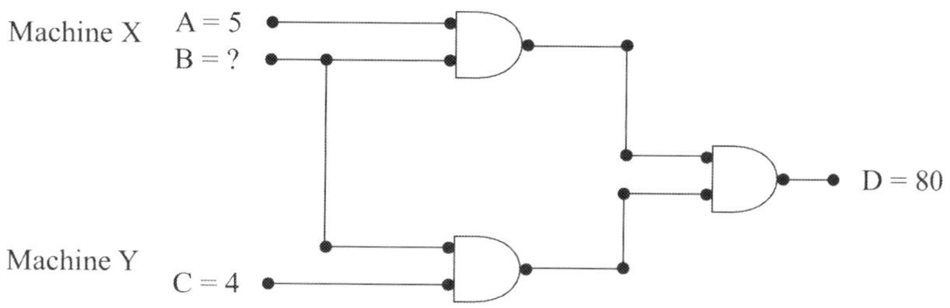

(A) 10 (B) 2

(C) 4 (D) 20

44. Simplify the expression -10(-2r + 3) + 3(1 - r)

(A) 17r - 27 (B) 4

(C) -6 - 4r (D) -6

45. Simplify the expression -2(1 + 17v) - 14v

(A) -107v + 140 (B) -2 - 48v

(C) 127v - 11 (D) -50v - 198

46. Simplify the expression and plot the graph

-115 >= 10b - 5

(A) $b \leq -23$: 

(B) $b \leq -46$: 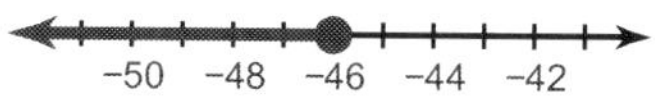

(C) $b \leq 6$: 

(D) $b \leq -11$: 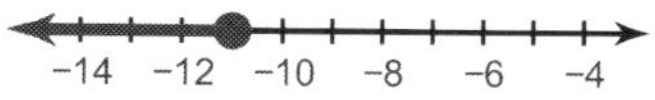

47. Simplify the inequality and graph the solution

$$-3 + 3x \geq 33 \text{ or } 10x + 2 \leq -98$$

(A) $x < 10$:

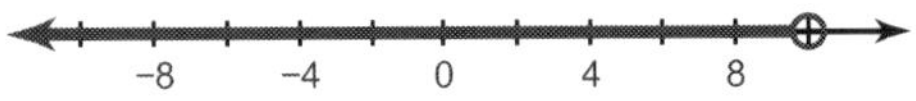

(B) $x \leq -4$ or $x > 1$:

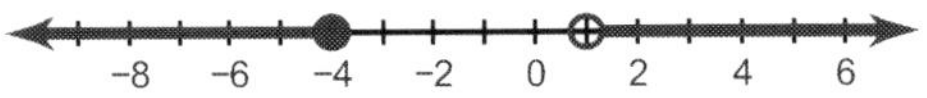

(C) $x \geq 12$ or $x \leq -10$:

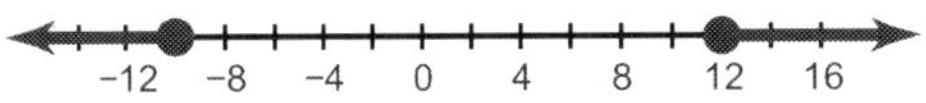

(D) $x > 8$:

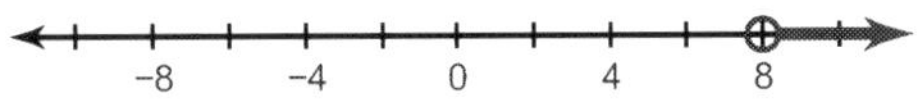

48. Which of the below represents all possible unique factors of 55

(A) 1,5,11,18,55

(B) 1,5,11,,33,55

(C) 1,5,11,55

(D) 1,11,55

49. Find the value of x from the figure given below

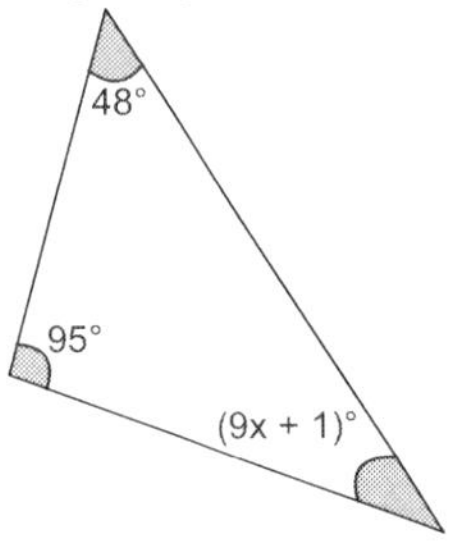

(A) 3

(B) 2

(C) 9

(D) 4

50. Find the value of x from the figure given below

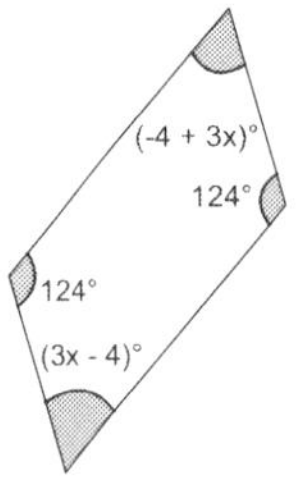

(A) 10

(B) 20

(C) 25

(D) 16

51. Find the value of x from the figure given below

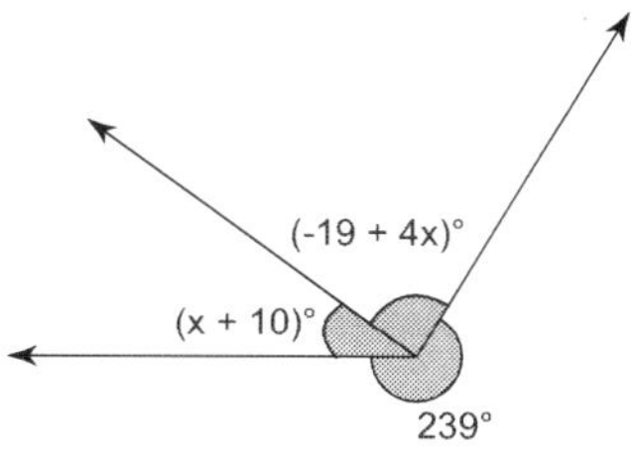

(A) -2
(B) 7
(C) 26
(D) 17

52. Solve the inequality and graph its solution

$$7 + |5x + 7| \geq 9$$

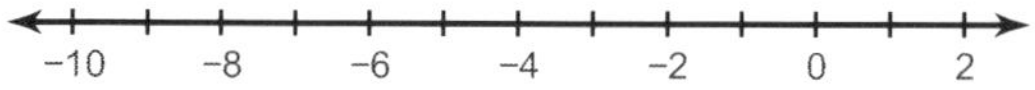

53. Find the midpoint of the line segment $\overline{RS}$ given below

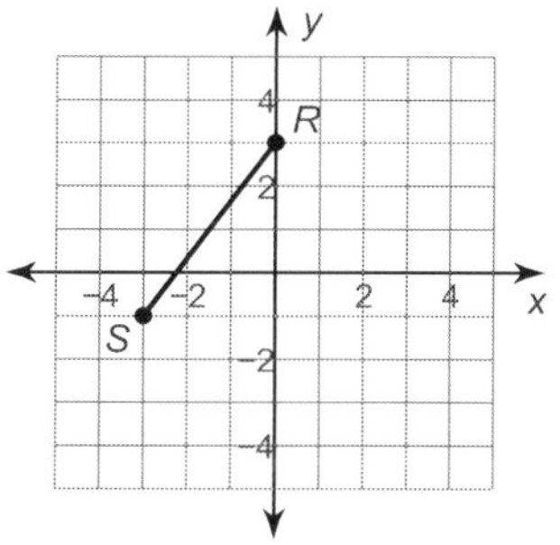

(A) (-6,-5)
(B) (1.5,2)
(C) (1.5,-2)
(D) (-1.5,1)

54. A painting is 14.7 in wide and 16.8 in tall. If it is reduced to a height of 2.4 in, then how wide will it be?

(A) 1.9 in
(B) 2 in
(C) 0.3 in
(D) 2.1 in

55. Households consume much more electricity when the weather is warmer:

Temperature (°F	Electricity (kWh)
58	32.4
61	34.3
63	34.2
71	39.2
78	42
85	44.8

This can be modeled by the equation $y = 0.466x + 5.51$ where x is the average daily temperature in °F and y is the average amount of electricity consumed in kilowatt-hours (kWh).

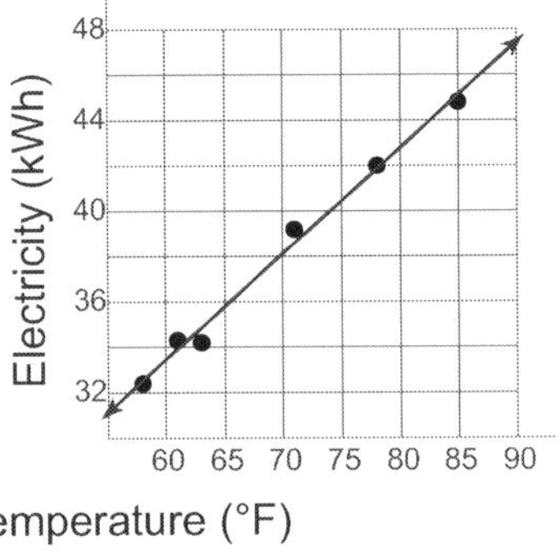

What temperature would it need to reach in order for 38 kWh to be consumed? Round your answer

to the nearest degree.

(A) 87°F

(B) 80°F

(C) 61°F

(D) 70°F

56. Evaluate $-2 - -2\frac{1}{2} - 3\frac{3}{4}$

SS High school conducted a survey on 6800 students for their favorite Soft Drinks. The below pie chart represents percentage of student votes. Answer the questions 7 - 12.

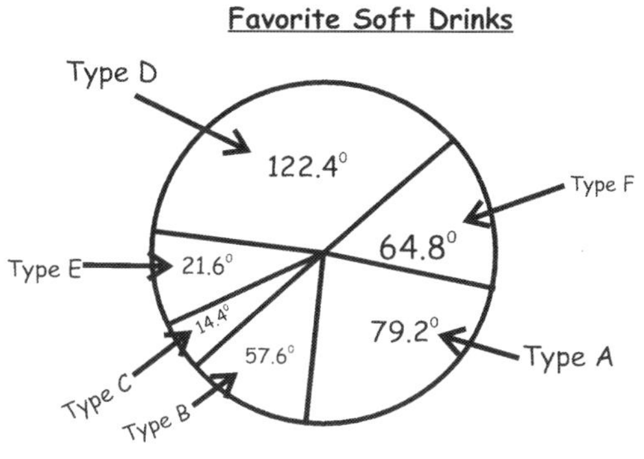

57. Find the percentage of the votes received together for types E and F

(A)	18%	(B)	20%
(C)	26%	(D)	24%

58. How many more votes were received for type C when compared to type D ?

(A)	2754	(B)	2040
(C)	2018	(D)	2900

SHSAT Practice Test 6

Evaluate the below expressions.

1. Evaluate $h - h + 4 + j^2$; where $h = -3$, and $j = 7$

(A) 49 (B) 53

(C) 59 (D) 46

2. Evaluate $|-7x - 9| = 16$, and find the sum of the solutions.

(A) $\frac{18}{7}$ (B) $\frac{9}{2}$

(C) $\frac{-7}{3}$ (D) $\frac{-18}{7}$

3. If $64^{-p+2} \cdot 16 = 16$

(A) -10 (B) -2

(C) 6 (D) 2

4. Find $x + y^2$ if $16x - 13y = -143$ and $3x + 13y = -104$

(A) 25 (B) 38

(C) -12 (D) 12

5. A straight line of slope $-\frac{5}{6}$ is passing through the points $A(-7, 6)$ and $B(5, y)$. Find the value of x.

(A) -9 (B) -4

(C) -2 (D) 2

6. A shop is offering 20% discount for one day on a video game that is priced at \$9.99? What is the selling price of the video game after discount ?

(A) \$6.10 (B) \$7.99

(C) \$7.68 (D) \$7.20

7. Evaluate $\frac{{}^{21}C_4}{3} = ?$

(A) 1654 (B) 1125

(C) 1490 (D) 1995

8. The original price of a chocolate bar is $2.20 and a 1% sales tax is added to the price. Find the sale price of the chocolate bar ?

(A) $2.10 (B) $2.90

(C) $0.22 (D) $2.22

9. Evaluate $-3 + {}^{25}C_3$

(A) 3,841 (B) 2,297

(C) 2,043 (D) 4,112

10. Evaluate $5 + {}^{18}P_5$

(A) 1028165 (B) 1128160

(C) 153165 (D) 958165

11. John bought 2 adult tickets and one child ticket at Fairfax carnival for $30.50. Lucy bought 3 adult tickets and one child ticket at Fairfax carnival for $40.40. What is the price of the adult ticket ?

(A) $10.45 (B) $8.90

(C) $9.90 (D) $6.10

12. Jack has the rope of length $205.8073 \text{ X } 10^{-2}$. He needs a rope of length 2172.67×10^{-3}. How much more rope does he need to buy ?

(A) 114.596×10^{-6} (B) 114.597×10^{-6}

(C) 114.597×10^{-3} (D) 1145.975×10^{-3}

13. If 8 ft long iron rod of uniform thickness weighs 28 lb, what will be the weight of 10 ft long iron rod of the same thickness ?

(A) 34.8 (B) 35.9

(C) 35 (D) 36

14. The sum of two numbers is 30 and their difference is eight. Find the product of the numbers ?

(A) 209 (B) 201

(C) 309 (D) 219

15. Which of the below graph best represents the function

$$f(x) = x^2 + 9x + 18$$

(A)

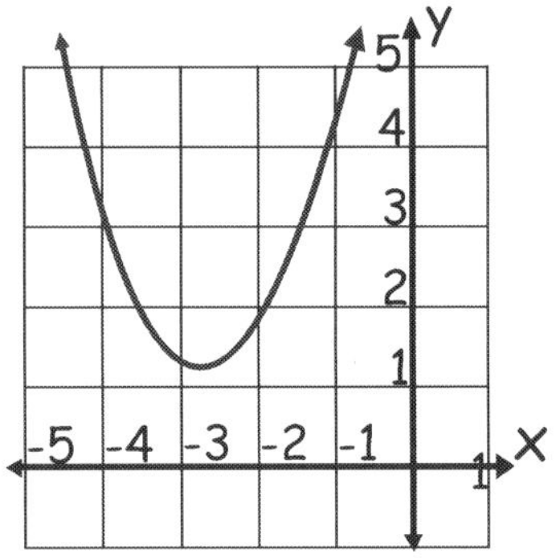

(B)

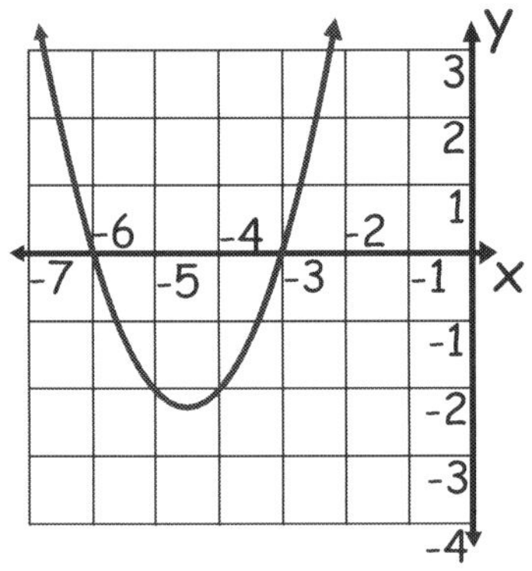

(C)

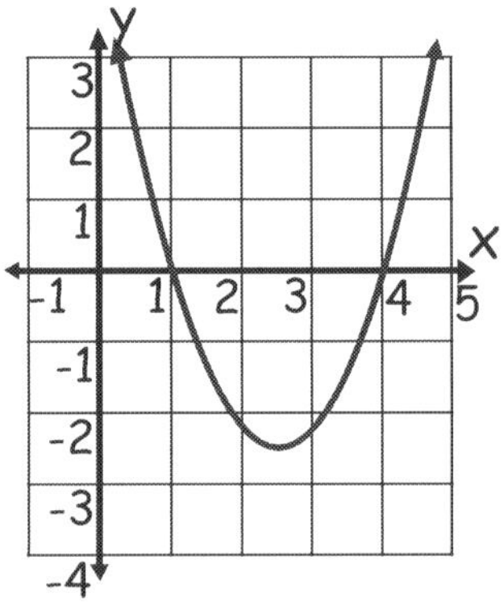

(D)

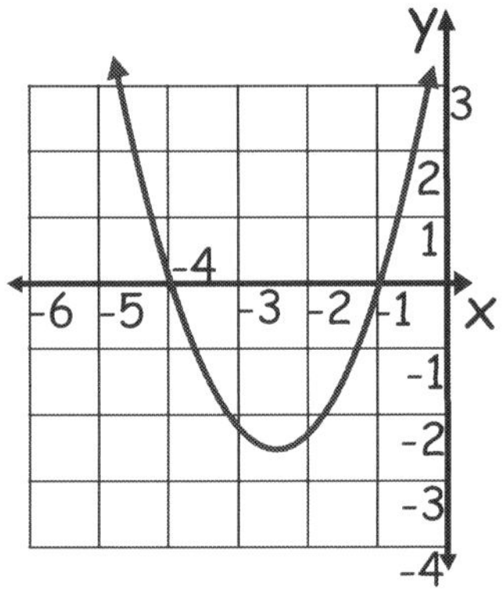

16. Three colored boxes Red, Green and Blue, together have 108 balls. There are twice as many balls in the green and red boxes combined as there are in the blue box and twice as many in the blue box as there are in the red box. How many balls are there in the green box ?

(A) 18 (B) 36

(C) 45 (D) 54

17. A number consists of two digits whose sum is 11. If 27 is added to the number, then the digits change their places. Find the number ?

(A) 47 (B) 65

(C) 83 (D) 92

18. Roma purchased a Furniture set at \$8,400.00 and sells it at a 80% price increase. What is the selling price of the Furniture set ?

(A) \$15,120.00 (B) \$7,980.00

(C) \$6,720.00 (D) \$1,680.00

19. The circumference of a circle is half of the perimeter of a rectangle. The area of circle is 1386 sq.m. What is the area of the rectangle if 50 m is its width ?

(A) 4100 sq.m (B) 4353 sq.m

(C) 3600 sq.m (D) 4900 sq.m

20. Find the percentage change from 76 cm to 12 cm.

(A) 15.8% decrease (B) 133.3% increase

(C) 69.5% decrease (D) 84.2% decrease

21. Jade took a loan of \$21,100 at 14% interest for 2 years. Find the amount she needs to pay at the end of the loan term.

(A) \$27,026.39 (B) \$27,008.00

(B) \$24,054.00 (D) \$27,421.56

22. John took a loan of \$870 at 4% interest compounded semi annually for 5 years. Find the amount he needs to pay at the end of the loan term.

(A) \$1,060.53 (B) \$1,044.00

(C) \$1,287.81 (D) \$1,058.49

23. A farmhouse shelters 23 animals. Some are pigs and some are chickens. Altogether there are 84 legs. How many of each animal are there ? How many pigs are there in farmhouse ?

(A) 19 pigs (B) 2 pigs

(C) 2 pigs (D) 3 pigs

24. Find the missing terms in the below sequences

..., -5, ___, ___, -23, ...

(A) -11, -17
(B) -6, -9
(C) -8, -13
(D) -10, -15

25. Find the missing terms in the below sequences

..., -23, ___, ___, ___, ___, ___, 577, ...

(A) 75, 173, 271, 369, 467
(B) 173, 271, 369, 467, 565
(C) 77, 177, 277, 377, 477
(D) 76, 175, 274, 373, 472

26. Evaluate $(-2) \times (-6) + 5 - 5$

(A) 12
(B) 21
(C) 20
(D) 19

27. Solve the proportion $\frac{7}{10} = \frac{x-14}{x}$

(A) -3
(B) 2
(C) 46.67
(D) 5

28. If $P(A) = \frac{13}{20}$ $P(A \text{ and } B) = \frac{13}{50}$ *then* $P(B) = ?$

(A) $\frac{3}{20}$
(B) $\frac{2}{5}$
(C) $\frac{1}{2}$
(D) $\frac{33}{80}$

29. An acid solution was made by mixing 4 fl.oz. of a 58% acid solution and 2 fl. oz. of a 28% acid solution. What is the concentration of the mixture?

(A) 41% (B) 10%

(C) 55% (D) 48%

30. Ship X leaves the Pier two hours before ship Y. They travel in opposite directions. Ship Y travels at a speed of 10 mi/h for two hours, at this time both ships are 120 mi apart Find the speed of ship X.

(A) 30 mi/h (B) 10 mi/h

(C) 16 mi/h (D) 25 mi/h

31. Aysha can pick forty piles of strawberries in 11 hours. Misha can do the same in 14 hours. How long will it take them if they worked together ?

(A) 4.87 hours (B) 6.16 hours

(C) 7.5 hours (D) 6.79 hours

32. A bag contains five red marbles and seven blue marbles. Another bag contains five green marbles and five yellow marbles. Jade randomly picks one marble from each bag. Find the probability of picking a blue and a yellow marble ?

(A) $\frac{14}{33}$ (B) $\frac{7}{24}$

(C) $\frac{1}{16}$ (D) $\frac{1}{36}$

33. A magic bag contains five puzzles and eight video games. Henry picks up an item and then another without returning the former. Find the probability of him picking puzzles in both times ?

(A) $\frac{24}{91}$ (B) $\frac{5}{39}$

(C) $\frac{4}{15}$ (D) $\frac{1}{6}$

34. If 33 men can do a piece of work in 15 days by working for 8 hours then find in how many days 44 men can do the same work by working 6 hours a day?

(A) 15 days (B) 18 days

(C) 20 days (D) 12 days

35. Siya, Tom and Rita invested in the ratio of 4: 7: 9. After 6 months, Siya invested \$ 5000 more and after 2 months, Rita withdraws \$ 7500. Find the initial investment of Tom, if the share of Siya, Tom and Rita is in the ratio of 17: 28: 35 ?

(A) \$ 70000 (B) \$ 35000

(C) \$ 48000 (D) \$ 66000

36. Two machines X and Y takes 2 inputs and produce one output which is greatest of all. Machine Z takes the outputs of machines X and Y as inputs. Machine Z multiplies its inputs and produces it as ouput. Based on the given information and the below figure find the missing value ?

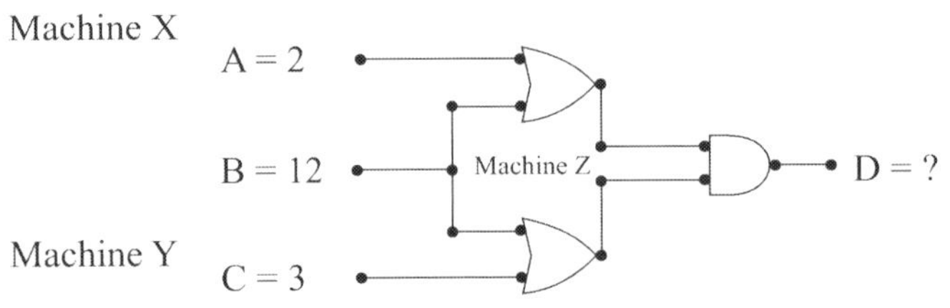

(A) 12 (B) 24

(C) 17 (D) 144

37. Evaluate the below series

$$a_1 = -3 \text{ , } a_8 = 384 \text{ , } r = -2$$

(A) -1 (B) 221

(C) 255 (D) 230

38. Find the number of terms in the series given below

$$\sum_{i=1}^{n} (8m + 2) = 648$$

(A) 15 (B) 10

(C) 12 (D) 13

39. Find the number of terms in the series given below

$$\sum_{i=1}^{9} \left(\frac{-2}{3}\right)^{i-1}$$

(A) $\frac{4039}{6563}$

(B) $\frac{4036}{6563}$

(C) $\frac{4039}{6561}$

(D) $\frac{3}{5}$

40. How many unique permutations of the letters of the word **EMPATHY** can be formed ?

(A) 5040

(B) 2640

(C) 5420

(D) 3540

41. The figure below represents a picture frame and the shaded regions represent the actual pictures. The remaining part is the border around. If all the line segments in the figure are either horizontal or vertical and the picture are of same size. find the dimensions of the picture inside ?

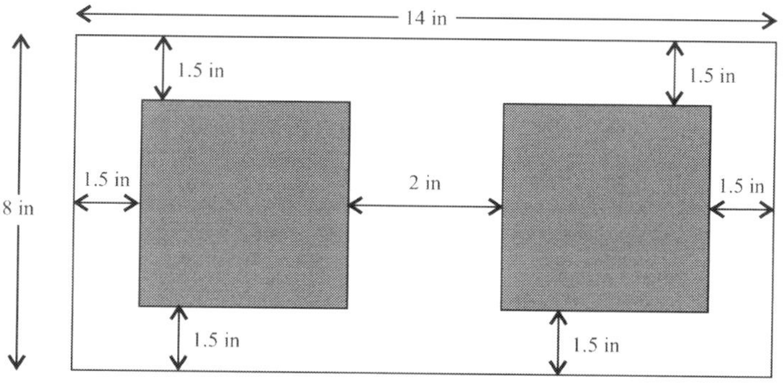

(A) 4.5 in , 5 in

(B) 4.5 in , 6.5 in

(C) 5 in , 5.5 in

(D) 5 in , 9 in

42. For positive numbers n , n + 1 , n + 2 , n + 4 and n + 8. The mean is how much greater than the median ?

(A) 0

(B) 1

(C) n + 1

(D) n + 2

43. In how many ways 8 players can be selected from a group of 12 players ?

(A) 495

(B) 340

(C) 120

(D) 520

44. Simplify the expression -(5k + 4) + 4(4k + 7)

(A) 12k + 24 (B) 12k + 29

(C) -17 + 5k (D) 11k + 24

45. Simplify the expression -3(-5 + a) + 11a

(A) 15 + 8a (B) 33 + 8a

(C) -1 + 8a (D) 20 + 8a

46. Simplify the expression and plot the graph $11 > 7 + \frac{v}{2}$

(A) v < 8 :

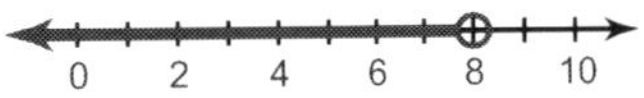

(B) v > -21 :

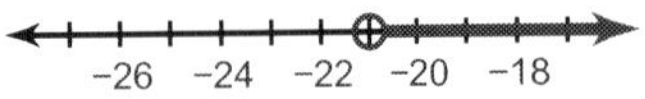

(C) v > 8 :

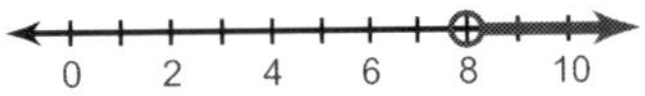

(D) v > 8 :

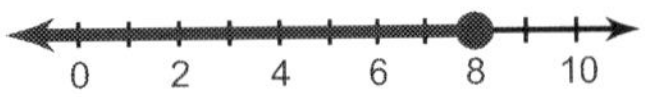

47. Simplify the inequality and graph the solution

-54 < 8a - 6 < 42

(A) -10 < a ≤ 5 :

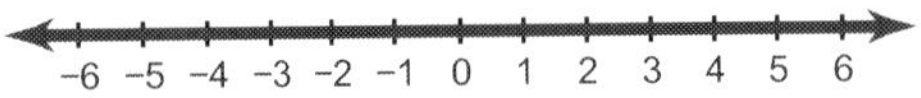

(B) 11 ≤ a < 12 :

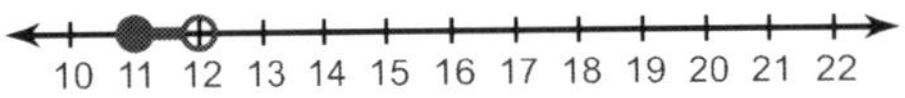

(C) No solution :

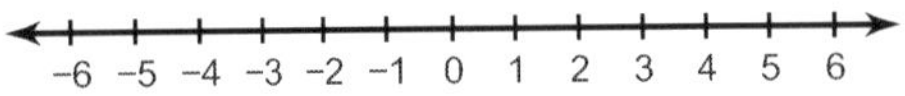

(D) -6 < a <6 :

48. Which of the below represents all possible unique factors of 74

(A) 1,74 (B) 1,37,45,74

(C) 1,2,36,37,74 (D) 1,2,37,74

49. Find the value of x from the figure given below

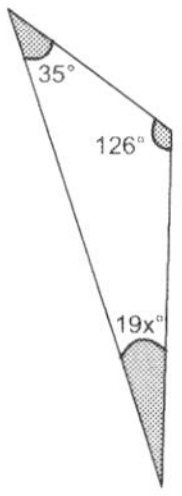

(A) 1 (B) 4

(C) 6 (D) 11

50. Find the value of x from the figure given below

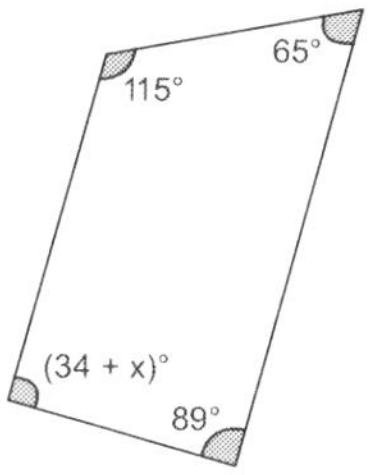

(A) 41 (B) 46

(C) 57 (D) 49

51. Find the value of x from the figure given below

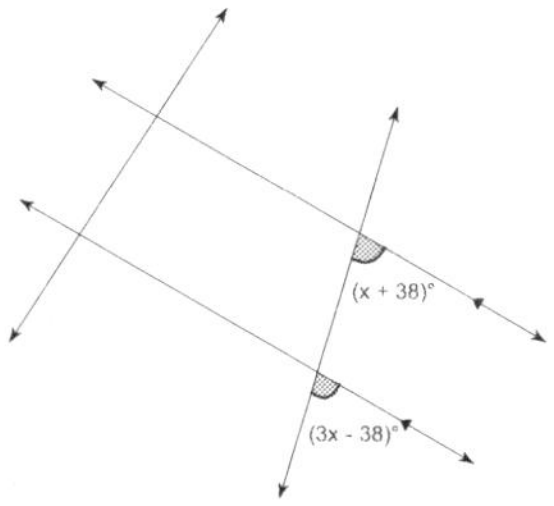

(A) 38 (B) 35

(C) 55 (D) 16

52. Solve the inequality and graph its solution

$$10\left|-2x+2\right| \leq 20$$

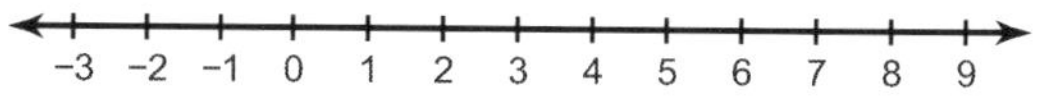

53. Find the midpoint of the line segment $\overline{RS}$ given below

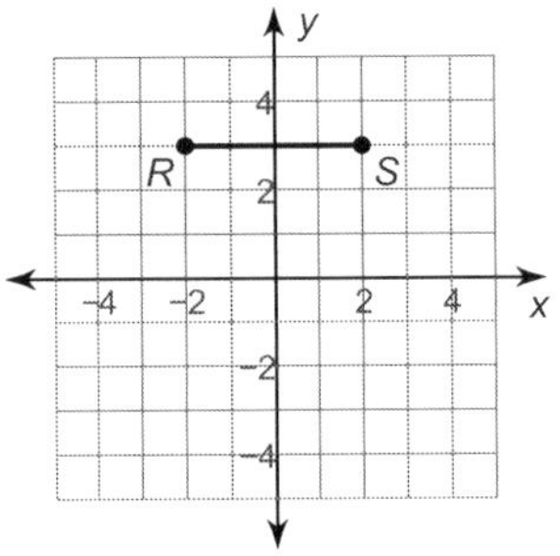

(A) (2.5,0.5) (B) (0,3)

(C) (2,0) (D) (-6,-3)

54. A frame is 2.4 in tall and 3.5 in wide. If it is enlarged to a height of 4.8 in, then how wide will it be?

(A) 9.6 in (B) 2.4 in

(C) 7 in (D) 7.6 in

55. The Hurricane Hunters took the following measurements from a hurricane over several days as it developed:

Air Pressure (kP	Wind Speed (knots)
924	128
939	112
960	77.7
981	60
996	42.2
1,007	32.1

They found that the air pressure and wind speed are related in the following way: $y = -1.17x + 1210$ where x is the pressure in millibars (kPa) and y is the maximum sustained wind speed in knots (nautical miles per hour).

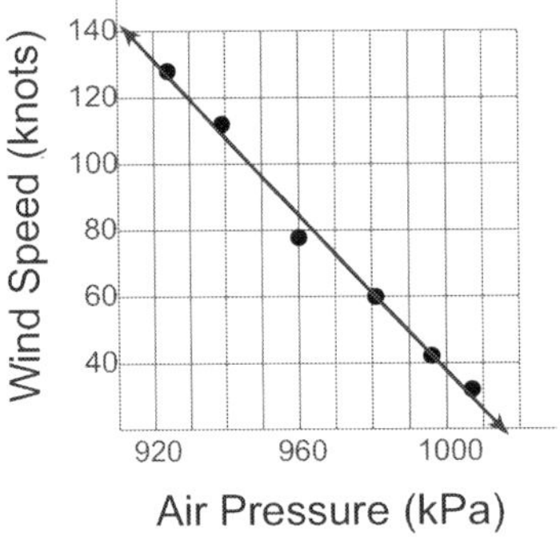

The model indicates that a wind speed of 100 knots is associated with what air pressure?

(A) 949 kPa (B) 966 kPa

(C) 1004 kPa (D) 741 kPa

56. Evaluate $-\frac{3}{2}+4\frac{1}{4}+-1\frac{1}{4}$

SS High school conducted a survey on 6800 students for their favorite Soft Drinks. The below pie chart represents percentage of student votes. Answer the questions 7 - 12.

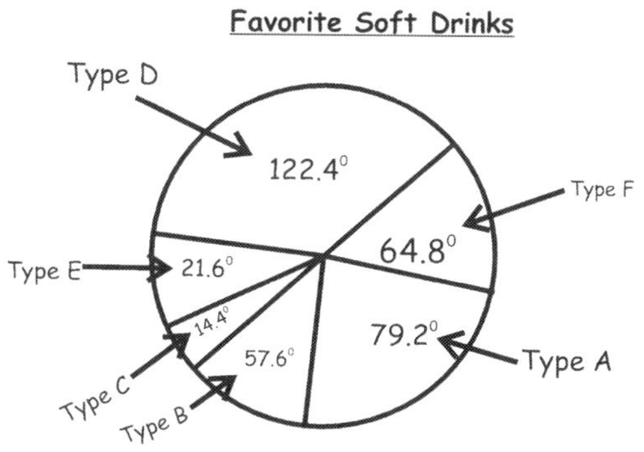

57. How many votes were received for types B and E together ?

(A) 1496 (B) 1420

(C) 1318 (D) 1485

58. How many votes were received for types B and C together ?

(A) 1290 (B) 1366

(C) 1360 (D) 1450

SHSAT Practice Test 7

1. Dan can do 1/3 of the work in 5 days, Sara can do 2/5 of the work in 10 days. In how many days can both of them together complete the work?

(A) $7\frac{3}{4}$ days (B) $8\frac{5}{6}$ days

(C) $10\frac{7}{9}$ days (D) $9\frac{3}{8}$ days

2. Evaluate $|9x - 4| = 4$, and find the sum of the solutions

(A) 8 (B) $-\frac{1}{9}$

(C) $\frac{8}{9}$ (D) 7

3. Simplify $\frac{16}{\left(\frac{1}{64}\right)^{-n}} = \frac{1}{4}$. Find the value of 'n'?

(A) 1 (B) 4

(C) $\frac{3}{2}$ (D) 12

4. Find $(x - y)^2$ if $4x + 15y = -120$ and $11x - 15y = -105$

(A) 11 (B) -11

(C) 124 (D) 121

5. In how many ways 3 sales jobs can be filled from the 45 applications received for it ?

(A) 42,570 (B) 85,140

(C) 20,770 (D) 14,190

6. Find the equation of a straight line passing through the pints G(4, −4) and perpendicular to $y = \frac{4}{7}x - 3$

(A) $y = -\frac{1}{2}x + 3$ (B) $y = -\frac{7}{4}x + 3$

(C) $y = -\frac{5}{4}x + 3$ (D) $y = \frac{7}{4}x + 3$

7. Albert purchased a Scrabble game at \$20.50 and sells it at a 20% price increase. Find the selling price of the Scrabble game.

(A) \$4.10 (B) \$24.60

(C) \$17.43 (D) \$16.40

8. A minute cell divides into 5 parts every 3 minutes. In 30 minutes how many cells are formed ?

(A) 3^{10} (B) 5^{9}

(C) 5×10 (D) 5^{10}

9. Best Brand store is selling TV at $362.70 which was originally priced as $604.50. What is percentage of discount offered ?

(A) 40% (B) 15%

(C) 50% (D) 38%

10. Sia has 5 dresses, 6 pairs of shoes and 5 hand bags. In how many ways can she dress for the party ?

(A) 150 (B) 30

(C) 250 (D) 16

11. Larry can solve forty puzzles in 15 hours. Mary can solve the same forty puzzles in 9 hours. Together they can solve the 40 puzzles in how many hours ?

(A) 5.01 hours (B) 6.63 hours

(C) 5.63 hours (D) 5.53 hours

12. A tiles company charges $10 per square foot. Raman wants to do tiles in four rooms of equal area in his office. If each room has an area of 140 sq ft. How much does it cost in total ?

(A) $400 (B) $560

(C) $550 (D) $5600

13. Ritvik mixes 2 Spoons of Chocolate powder for every 20 oz of milk that he drinks. If he drinks 360 oz of milk in a month, how many spoons of chocolate powder does he need for a month ?

(A) 10 (B) 36

(C) 18 (D) 72

14. Rik pulls a card from the deck of cards. What is the probability that it is a spade ?

(A) $\frac{1}{4}$ (B) $\frac{1}{52}$

(C) $\frac{1}{13}$ (D) $\frac{2}{52}$

15. When two dice are rolled. what is the probability that the sum is greater than 4 ?

(A) $\frac{1}{6}$ (B) $\frac{5}{9}$

(C) $\frac{2}{9}$ (D) $\frac{5}{6}$

16. Two bus tickets from city A to B and three tickets from city A to C cost \$77 but three tickets from city A to B and two tickets from city A to C cost \$73. What are the fares for cities B and C from A respectively ?

(A) \$ 4, \$ 23 (B) \$ 13, \$ 17

(C) \$ 15, \$ 14 (D) \$ 17, \$ 13

17. The simple interest accrued on an amount of \$ 15,000 at the end of two years is \$2400. At the same interest rate, what will be the corresponding Compound Interest ?

(A) \$ 2496 (B) \$ 2512

(C) \$ 2538 (D) \$ 2474

18. Cathy purchased a bag priced at \$37.95. She received a discount of 40% at the billing counter as a preferred customer. What is the selling price of the bag ?

(A) \$53.13 (B) \$22.77

(C) \$36.05 (D) \$15.18

19. Jill selected a X - Box for his son that is tagged with price \$499.95. While paying the bill, he noticed that 6% tax is added to the tagged price. He has a store credit of \$50. How much did he pay at the counter for the purchase of X - Box ?

(A) \$529.90 (B) \$322.77

(C) \$479.95 (D) \$630.25

20. Find the percentage change from 50 ft to 99 ft ?

(A) 198% increase (B) 49.5% decrease

(C) 98% increase (D) 26.5% increase

21. Johnson gave a personal loan to Dan an amount of \$990 at 15% interest. If Dan repays it after 2 years, how much amount did Johnson received ?

(A) \$319.27 (B) \$1,309.27

(C) \$1,287.00 (D) \$148.50

22. Brooks gave a personal loan to Kylo an amount of \$17,000 at 6% interest. compounded annually. If Kylo repays it after 2years, how much amount will Brooks receive ?

(A) \$19,081.63 (B) \$19,111.79

(C) \$19,101.20 (D) \$19,118.11

23. There are 12 animals in the barn. Some are ducks and some are horses. There are 34 legs in all. How many of each animal are there ? How many ducks are there in the barn ?

(A) 8 ducks (B) 9 ducks
(C) 7 ducks (D) 10 ducks

24. Find the missing terms in the below sequences

..., -22, ___, ___, -4, ...

(A) -17, -12 (B) -16, -10
(C) -19, -16 (D) -18, -14

25. Evaluate $((-8)-(6-7))\times(-8)$

(A) 49 (B) 56
(C) 65 (D) -50

26. Find the missing terms in the below sequences

..., 4, ___, 36, ...

(A) 12 (B) 4
(C) 16 (D) 1

27. Solve the proportion $\frac{x+3}{x}=\frac{3}{6}$

(A) -1.26 (B) -10.2
(C) -6 (D) 5.2

28. If $P(A)=\frac{3}{10}$ $P(A \text{ and } B)=\frac{3}{25}$ *then* $P(B|A)=$?

(A) $\frac{4}{25}$ (B) $\frac{2}{5}$
(C) $\frac{3}{10}$ (D) $\frac{21}{200}$

29. Scott mixed together 5 gal. of Brand A fruit drink and 3 gal. of apple juice. Find the percent of fruit juice in Brand A if the mixture contained 50% fruit juice.

(A) 32%
(B) 12%
(C) 20%
(D) 35%

30. Eric started from home to Washington DC one hour before Sam left to Florida. They drove in opposite directions. Sam drove at 75 mi/h for three hours, at this time they are 485 miles apart. Find Eric's average speed.

(A) 65 mi/h
(B) 55 mi/h
(C) 70 mi/h
(D) 30 mi/h

31. Pam takes 15 hours to pack forty boxes of books. Kate can do the same in 12 hours. How long will it take if they work together ?

(A) 8.07 hours
(B) 7.3 hours
(C) 7.21 hours
(D) 6.67 hours

32. A box contains five nickels and eight dimes. Marlo picks a coin and keeps it back in the box and shuffles them. Find the probability of him picking up a nickle for the first time , a dime in the second try and a nickle in the last try ?

(A) $\frac{1}{4}$
(B) $\frac{1}{16}$
(C) $\frac{200}{2197}$
(D) $\frac{4}{15}$

33. Company X has 237 local offices and one national office and one sales office. If each local office has two CEO's and each of the other two offices has 4 CEO's how many CEO's does the company have all together ?

(A) 482
(B) 476
(C) 239
(D) 225

34. Evaluate the sum of the below series

$$a_1 = -4 \text{ , } a_{10} = -2048 \text{ , } r = 2$$

(A) -4092
(B) 4844
(C) -3777
(D) -2792

35. Bella has thirteen shirts in her closet, four blue, four green, and five red. She randomly select a shirt each day. Find the probability of her wearing blue shirt on Monday, Tuesday, and Wednesday ?

(A) $\frac{7}{52}$ (B) $\frac{1}{32}$

(C) $\frac{10}{91}$ (D) $\frac{2}{143}$

36. A juice carton has 160 litres of Orange and Apple juice. The ratio of Orange to Apple juices is 5: 3. 40 liter of mixtures is taken out and replaced with Orange juice. Find the new ratio of Orange to Apple juices.

(A) 9 : 25 (B) 23 : 9

(C) 9 : 23 (D) 25 : 9

37. In Grade 8 class of RC Middle school, 3/5 of the students are girls and rest are boys. If 2/9 of the girls and 1/4 of the boys are absent today, what fraction of students are present today ?

(A) $\frac{17}{25}$ (B) $\frac{18}{49}$

(C) $\frac{23}{30}$ (D) $\frac{23}{36}$

38. Two machines X and Y takes 2 inputs and produces one output which is product of all inputs received. Machine Z takes the outputs of machines X and Y as inputs. Machine Z multiplies its inputs and produces it as ouput. Based on the given information and the below figure find the missing value ?

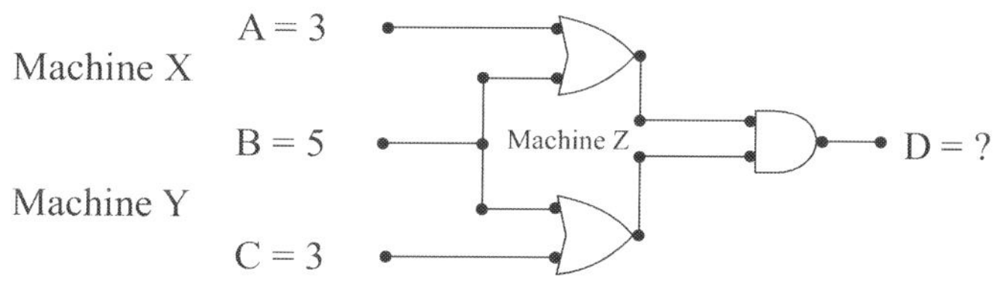

(A) 5 (B) 15

(C) 45 (D) 225

39. Find the number of terms in the series given below

$$\sum_{i=1}^{n} (9i - 10) = 5320$$

(A) 32 (B) 40

(C) 42 (D) 35

40. Average marks of Dan in two of his science tests is 87. How many marks does he need to score in the third test to get an average of 90 ?

(A) 92 (B) 90

(C) 96 (D) 87

41. A plane travelled 540 miles to Virginia from Dallas. It took 6 hours to go to Virginia. It took 18 hours to travel back to Dallas due to bad weather. Find the average speed of the plane.

(A) 35 mph (B) 60 mph

(C) 120 mph (D) 45 mph

42. The Tiny Tots school has two branches A and B. Branch A has twelve elementary class rooms and 10 middle school class rooms with 768 students in total. Branch B has four elementary class rooms and six middle school class rooms with a total of 416 students in total. Each elementary and middle school class rooms have same number of kids. How many students are there in each of them ?

(A) elementary class room : 14 , middle school class room : 60
(B) elementary class room : 14 , middle school class room : 55
(C) elementary class room : 35 , middle school class room : 45
(D) elementary class room : 60 , middle school class room : 14

43. The figure below represents three squares A , B , C with areas of 144 , 81 , 225 sq.units respectively. A triangle is formed when the squares are joined together in such a way each square touches other two squares at one point. If D is the area of the triangle then D = ?

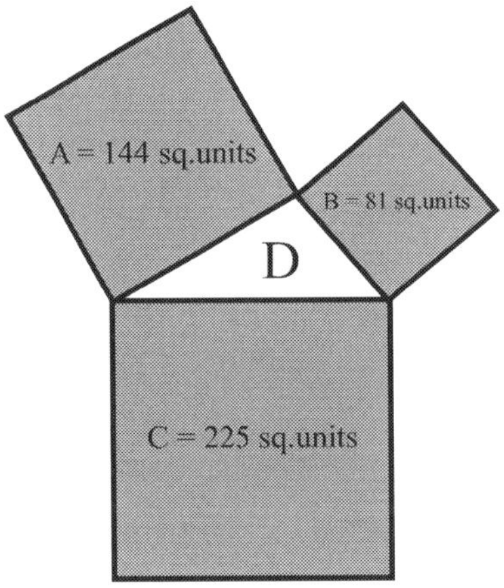

(A) 150 (B) 144

(C) 54 (D) 36

44. Simplify the expression 9(r + 10) + 5(r + 6)

(A) -12r - 98 (B) 10 - 8r

(C) -12r - 104 (D) 14r + 120

45. Simplify the expression -3(12 - 2k) - 6

(A) -100k + 305 (B) -100k + 293

(C) -42 + 6k (D) -42 + 21k

46. Simplify the expression and plot the graph $-6 < 6(7 + x)$

(A) $x > -12$: 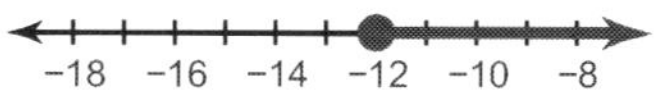

(B) $x > -12$: 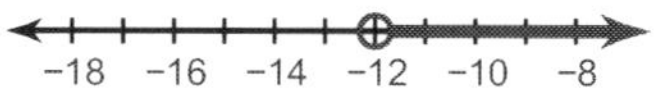

(C) $x < -12$: 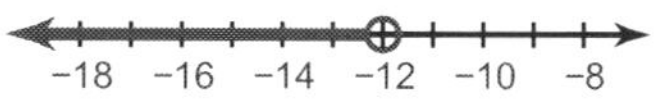

(D) $x > -8$: 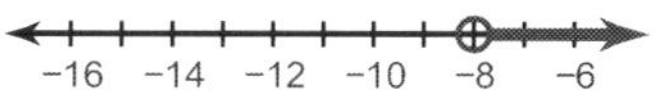

47. Simplify the inequality and graph the solution

$$2r - 6 > -10 \text{ or } 11r + 1 \leq -131$$

(A) $r \geq -1$: 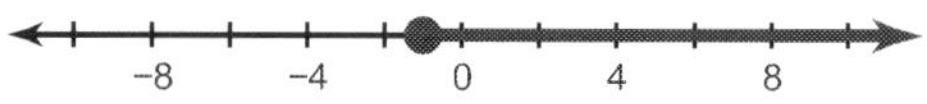

(B) $r \leq -12$: 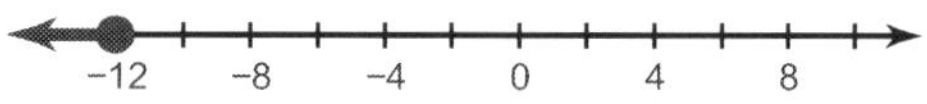

(C) $r > -2$ or $r \leq -12$: 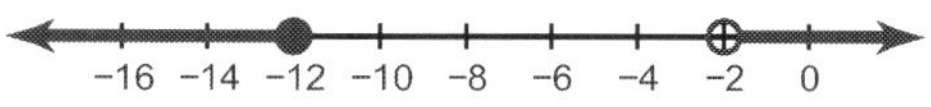

(D) $r \geq -1$ or $r < -11$: 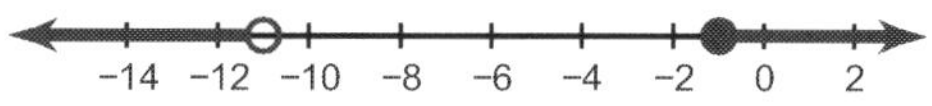

48. Which of the below represents all possible unique factors of 63

(A) 1,3,7,9,63 (B) 1,3,7,22,63

(C) 1,3,7,9,21,63 (C) 1,63

49. Find the value of x from the figure given below

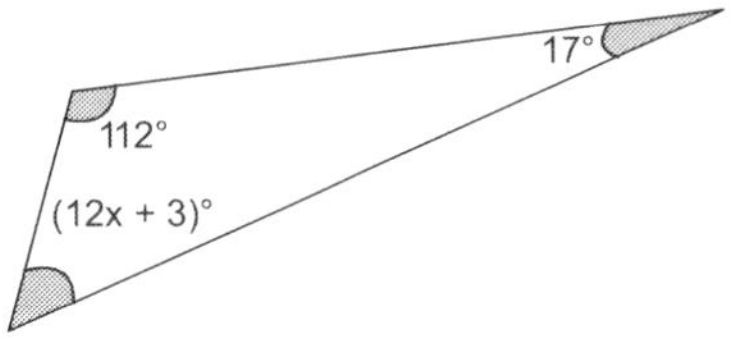

(A) 1 (B) 8

(C) 3 (D) 9

50. Find the value of x from the figure given below

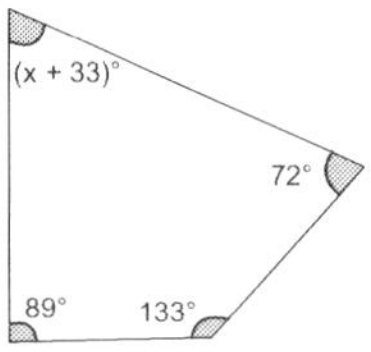

(A) 27 (B) 44

(C) 36 (D) 33

51. Find the value of x from the figure given below

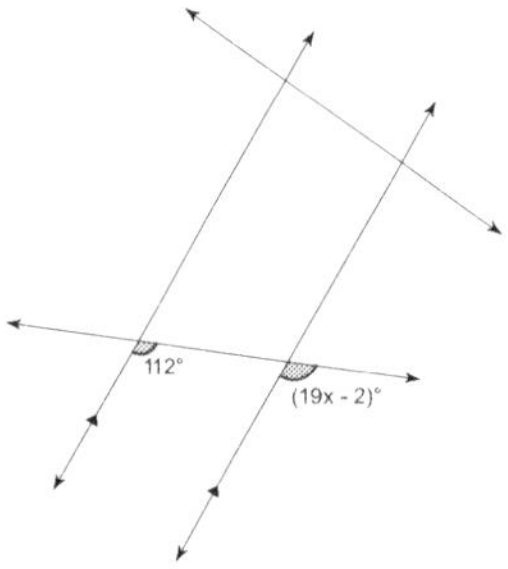

(A) 6 (B) -25

(C) -30 (D) -9

52. Solve the inequality and graph its solution

$$\frac{|2n+8|}{2} < 5$$

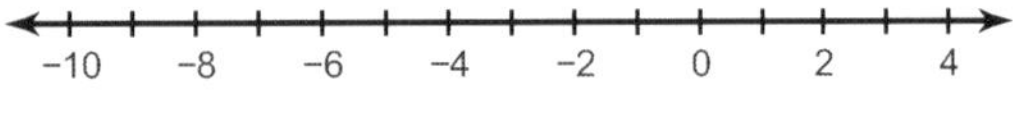

53. Find the midpoint of the line segment $\overline{ST}$ given below

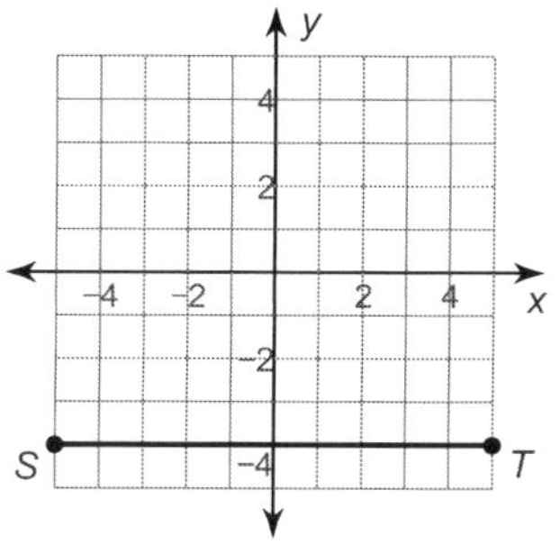

(A) (0,-4) (B) (-15,-4)

(C) (0.5,-4.5) (D) (5,0)

54. Maria reduced the size of a triangle to a width of 3.7 in. What is the new height if it was originally 14.8 in wide and 10.8 in tall?

(A) 0.9 in (B) 2.9 in

(C) 2.7 in (D) 43.2 in

55. Joe collects coins. Over a three-year period he collected 1,000 nickels. After organizing them by year, he found that the number of nickels from a given year was related to the number minted that year:

Minted (100 mill.)	In Sample
4.3	3
7.1	19
9.3	9
13.3	24
14.6	21
17.3	40

Joe discovered that this can be modeled by the equation $y = 2.28x - 5.66$ where x is the number of nickels minted in a particular year in hundreds of millions and y is the number of nickels from that year in his sample.

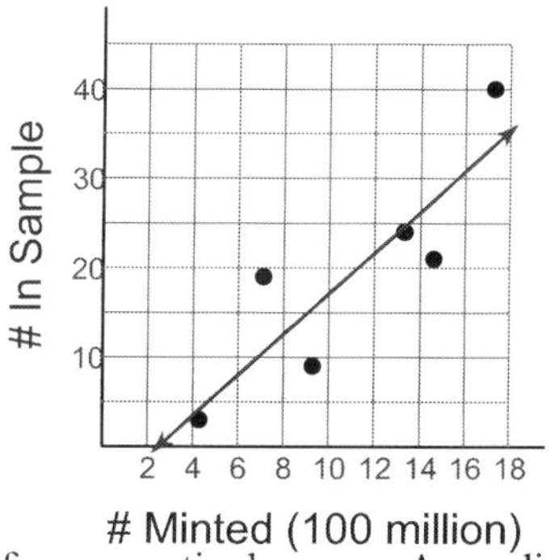

A sample of 1,000 nickels contains 20 from a particular year. According to the model, how many would you expect to have been minted that year? Round your answer to the nearest tenth of a hundred million.

(A) 9.6 hundred million nickels (B) 10.6 hundred million nickels

(C) 13 hundred million nickels (D) 11.3 hundred million nickels

56. Evaluate $\frac{4}{5} - \frac{1}{2} - \frac{1}{4}$

A happy living County spends the tax amounts received for various community development needs. The below pie chart represents the distribution of funds. Answer the questions 13 - 25.

County funds Allocation

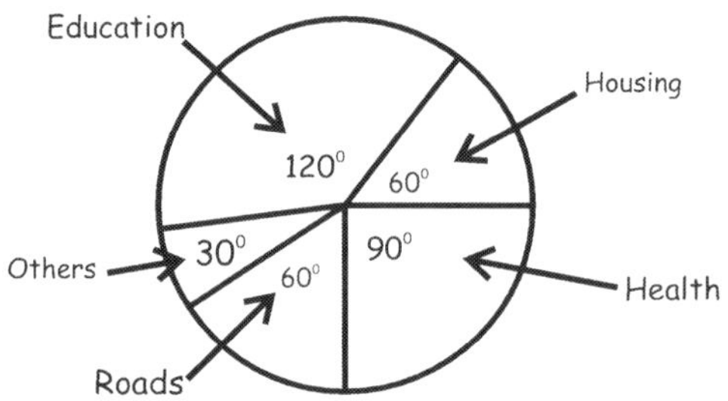

56. Find the ratio of the expenditure on education to health ?

(A) 3:2 (B) 2:3

(C) 3:4 (D) 4:3

57. Find the percentage of money was spent on housing ?

(A) 19 % (B) 16 %

(C) 15% (D) 16 %

Answer Keys

#1 Prime Factors

1. C
2. C
3. A
4. A
5. A
6. D
7. A
8. D

#2 Numerical Expressions

1. C
2. A
3. D
4. C
5. A
6. A
7. A
8. A

#3 Simplify Expressions

1. A
2. D
3. A
4. A
5. C
6. C
7. A
8. A

#4 Evaluate Expressions

1. C
2. A
3. C
4. A
5. B
6. A
7. D
8. A

#5 Exponential Expressions

1. C
2. D
3. B
4. D
5. B
6. C
7. C
8. D

#6 Distance Between 2 Points

1. A
2. B
3. A
4. B
5. B
6. D
7. C

#7 Midpoint

1. D
2. D
3. B
4. A
5. A
6. D
7. C
8. D

#8 Equation of a Straight line

1. C
2. B
3. C
4. D
5. B
6. D
7. A

#9 Slope (2 points)

1. A
2. B
3. B
4. C
5. A
6. C

#10 Slope intercept form

1. B
2. C
3. B
4. A
5. D
6. C
7. D

#11 Graph Slope

1. B
2. B
3. B
4. D
5. B
6. D
7. B
8. A

#12 Find the slope

1. A
2. A
3. A
4. D
5. A
6. D
7. C
8. C

#13 Parallel line slope

1. C
2. A
3. A
4. C
5. C
6. B
7. C
8. A

#14 Perpendicular line slope

1. A
2. B
3. B
4. D
5. B
6. A
7. A
8. A

#15 Radicals 1

1. C
2. A
3. D
4. B
5. B
6. B
7. A

#16 Radicals 2

1. B
2. C
3. A
4. C
5. B
6. B
7. C

#17 Inequalities

1. C
2. A
3. C
4. D
5. B
6. B
7. C
8. D

#18 One step word problems

1. A
2. D
3. A
4. A
5. D
6. B
7. B

#19 Circle area

1. D
2. A
3. D
4. B
5. A
6. C
7. C
8. C

#20 Volume of a Sphere

1. A
2. D
3. B
4. A
5. C
6. D
7. D
8. D

#21 Volume of rectangle,square,prisms

1. A
2. D
3. B
4. D
5. B
6. D
7. A
8. A

#22 Volume of cone cylinder

1. B
2. A
3. C
4. C
5. B
6. A
7. C

#23 Missing angle 1

1. B
2. D
3. D
4. A
5. A
6. C
7. C
8. B

#24 Missing angle 2

1. C
2. D
3. C
4. C
5. D
6. A
7. B
8. A

#25 Reflection

1. A
2. D
3. D
4. B
5. D
6. B
7. B
8. C

#26 Rotation

1. C
2. B
3. A
4. D
5. B
6. C
7. B
8. C

#27 Translation

1. B
2. C
3. B
4. B
5. A
6. A
7. D
8. C

SHSAT Practice Test 1

1. C
2. A
3. D
4. A
5. B
6. D
7. B
8. C
9. A
10. B
11. D
12. A
13. C
14. A
15. D
16. C
17. C
18. D
19. B
20. C
21. A
22. D
23. B
24. C
25. D
26. B
27. D
28. B
29. D
30. D
31. B
32. B
33. D
34. B
35. B
36. C
37. A
38. C
39. C
40. A
41. C
42. B
43. A
44. A
45. B
46. B
47. A
48. C
49. C
50. C
51. A
52. 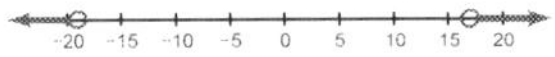

$n < -19$ or $n > 17$
53. B
54. A
55. A
56. $\frac{7}{4}$
57. B
58. C

SHSAT Practice Test 2

1. A
2. D
3. B
4. A
5. C
6. D
7. B
8. C
9. D
10. B
11. C
12. D
13. C
14. B
15. C
16. C
17. D
18. A
19. D
20. C
21. A
22. B
23. C
24. D
25. D
26. B
27. D
28. B
29. B
30. A
31. A
32. A
33. B
34. B
35. C
36. D
37. B
38. D
39. A
40. B
41. C
42. B
43. D
44. A
45. C
46. A
47. C
48. D
49. A
50. B
51. A
52. $|-3x-1|+8<10$

$-1 < x < \frac{1}{3}$

53. D
54. C
55. D
56. $-\frac{44}{15}$
57. D
58. A

SHSAT Practice Test 3

1. C
2. B
3. D
4. B
5. C
6. D
7. C
8. C
9. B
10. D
11. A
12. A
13. D
14. D
15. D
16. A
17. C
18. C
19. C
20. D
21. C
22. C
23. B
24. D
25. B
26. A
27. B
28. B
29. C
30. C
31. D
32. C
33. B
34. A
35. D
36. B
37. B
38. A
39. A
40. D
41. B
42. A
43. D
44. D
45. D
46. A
47. D
48. B
49. A
50. B
51. D
52. $6|3+8p| \le 30$

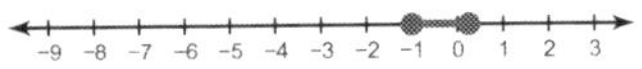

$-1 \le p \le \frac{1}{4}$

53. A
54. B
55. C
56. $-\frac{71}{42}$
57. B
58. D

SHSAT Practice Test 4

1. B
2. D
3. C
4. B
5. D
6. A
7. C
8. A
9. D
10. D
11. A
12. D
13. C
14. D
15. D
16. C
17. C
18. D
19. B
20. B
21. B
22. B
23. C
24. C
25. C
26. A
27. B
28. A
29. D
30. C
31. D
32. B
33. B
34. B
35. B
36. A
37. D
38. B
39. B
40. B
41. A
42. D
43. A
44. D
45. B
46. A
47. C
48. A
49. D
50. D
51. B
52. $|-7b-9|+7>72$

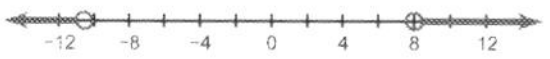

$b<-\frac{74}{7}$ or $b>8$

53. D
54. B
55. C
56. 0
57. C
58. A

SHSAT Practice Test 5

1. A
2. B
3. D
4. D
5. C
6. D
7. A
8. C
9. C
10. A
11. B
12. B
13. D
14. B
15. A
16. B
17. A
18. A
19. C
20. D
21. B
22. A
23. D
24. D
25. B
26. C
27. A
28. C
29. A
30. D
31. B
32. D
33. B
34. C
35. D
36. A
37. D
38. B
39. A
40. D
41. A
42. C
43. B
44. A
45. B
46. D
47. C
48. C
49. D
50. B
51. C
52. 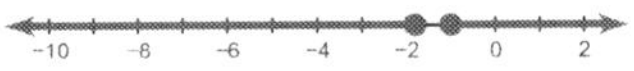

$x \geq -1$ or $x \leq -\frac{9}{5}$
53. D
54. D
55. D
56. $-\frac{13}{4}$
57. D
58. B

SHSAT Practice Test 6

1. B	16. D	31. B	46. A
2. D	17. A	32. B	47. D
3. D	18. A	33. B	48. D
4. D	19. A	34. A	49. A
5. B	20. D	35. A	50. C
6. B	21. B	36. D	51. A
7. D	22. A	37. C	52.  $0 \le x \le 2$
8. D	23. A	38. C	53. B
9. B	24. A	39. C	54. C
10. A	25. C	40. A	55. A
11. C	26. A	41. A	56. $-\frac{3}{2}$
12. C	27. C	42. B	57. A
13. C	28. B	43. A	58. C
14. A	29. D	44. D	
15. B	30. D	45. A	

SHSAT Practice Test 7

1.	D	16.	B	31.	D	46.	D
2.	C	17.	A	32.	C	47.	C
3.	A	18.	D	33.	A	48.	C
4.	D	19.	C	34.	A	49.	D
5.	D	20.	C	35.	D	50.	D
6.	B	21.	C	36.	B	51.	A
7.	B	22.	C	37.	C	52.	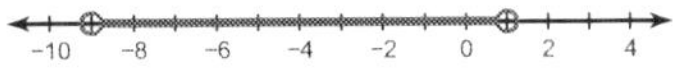 $-9 < n < 1$
8.	D	23.	C	38.	D	53.	A
9.	A	24.	B	39.	D	54.	C
10.	A	25.	B	40.	C	55.	D
11.	C	26.	A	41.	D	56.	$\frac{21}{20}$
12.	D	27.	C	42.	A	57.	D
13.	B	28.	B	43.	C	58.	D
14.	A	29.	A	44.	D		
15.	B	30.	A	45.	C		

Made in United States
Orlando, FL
05 November 2026

90229685R00131